PSYCHOPHYSICS

'This is an excellent introduction to the theory and practice of psychophysics. The clear text and accompanying Matlab routines make it a uniquely useful resource.'
N.E. Scott-Samuel, Experimental Psychology, University of Bristol, Bristol, UK

'For anyone who wants to know the way to do psychophysics, and wants to understand why it's done in this way, there is now a resource that's practical, thoughtful, and thorough. Kingdom and Prins have provided an exemplary guide to experimental practice. They lay out the methods for collecting and analyzing data, evaluate the up-sides and down-sides of the methods, and even include software to give the fledgling experimenter a running start. Yet the book isn't only for new-comers; many an old-timer will find a thing or two here to spur a re-thinking of old methodological habits. For any psychophysicist, from would-be to pro, this is a most useful book.'
Bart Farell, Institute for Sensory Research, Department of Biomedical & Chemical Engineering, Syracuse University and Departments of Ophthalmology & Physiology and Neuroscience, SUNY Upstate Medical University, Syracuse, NY, USA

'Kingdom and Prins have done an excellent job of combining a clear, logical and critical explanation of psychophysical techniques with practical examples. The result is a detailed, engaging and accessible text that is highly recommended for students new to the topic of psychophysics, and that provides an invaluable resource for advanced students, clinicians and research scientists alike.'
Benjamin Thompson, Department of Optometry and Vision Science, New Zealand National Eye Centre, University of Auckland

'This is an excellent, readable introduction to psychophysical methods. It covers all the basics from experimental methods, adaptive methods for setting stimulus levels, modern methods for analyzing the resulting data, and statistical methods for comparing alternative models of the data. I would recommend it heartily as a text for advanced undergraduate and beginning graduate students in any field in which psychophysical methods are used in behavioral experiments.'
Michael S. Landy, Department of Psychology, New York University, New York, NY, USA

'Kingdom and Prins have written a book that, though not exactly fun – let's face it, psychophysics isn't fun, it's hard – is clear and intelligent. If you want to know about psychophysics this would be a good place to start. If you know about psychophysics and want the tools to analyse your psychophysical experiments, then they have tools here aplenty with their 'Palamedes' toolbox of Matlab routines.'
Peter Thompson, Department of Psychology, University of York, UK

'The history of Psychophysics spans more than 150 years, but this very welcome new book by Kingdom & Prins is the first to consider systematically and in sufficient detail how modern psychophysics should be done. Psychophysical methods are carefully described and compared, and linked to mathematical theory that motivates different forms of data analysis and model fitting. A key feature is that the book is integrated with a software package (Palamedes) that enables readers to carry out sophisticated analyses and evaluations of their data that were impossible just a few years ago. Essential reading (and doing) for new and experienced researchers.'
Mark Georgeson, Vision Sciences, Aston University, Birmingham, UK

'The clear and logical structure of each chapter makes the book very useful for students coming to terms with some of the many dichotomies of psychophysics: Class A vs. Class B observations, or Type 1 and Type 2 experiments, for example. A student can be directed to read relevant sections of the book where they will find clear, worked through examples for each definition. At the end of many of the chapters are some useful exercises by which the student can assess their understanding of the concepts introduced in each chapter. I am sure this book will instantly become the book of choice for graduate-level courses in psychophysics.'
Laurence Harris, Centre for Vision Research, York University, Toronto, Canada

Companion Web Site:

http://books.elsevier.com/companions/9780123736567

Psychophysics: A Practical Introduction

Resources:

- **All figures from the book available**
- **Access to *Palamedes*, a new toolbox of MATLAB® routines for analyzing psychophysical data, created by the authors**

PSYCHOPHYSICS

A PRACTICAL INTRODUCTION

Frederick A.A. Kingdom
and
Nicolaas Prins

ELSEVIER

AMSTERDAM • BOSTON • HEIDELBERG • LONDON • NEW YORK • OXFORD •
PARIS • SAN DIEGO • SAN FRANCISCO • SINGAPORE • SYDNEY • TOKYO

Academic Press is an imprint of Elsevier

Academic Press is an imprint of Elsevier
32 Jamestown Road, London NW1 7BY, UK
30 Corporate Drive, Suite 400, Burlington, MA 01803, USA
525 B Street, Suite 1900, San Diego, CA 92101-4495, USA

First edition 2010

British Library Cataloguing-in-Publication Data
A catalogue record for this book is available from the British Library

Library of Congress Cataloging-in-Publication Data
A catalog record for this book is available from the Library of Congress

ISBN: 978-0-12-373656-7

For information on all Academic Press publications
visit our website at www.elsevierdirect.com

Typeset by Macmillan Publishing Solutions
www.macmillansolutions.com

Printed and bound in the United Kingdom
Transferred to Digital Print 2010

FK would like to dedicate this book to his late parents Tony and Joan, and present family Beverley and Leina. NP would like to dedicate this book to his mother Nel and late father Arie.

Contents

5. ADAPTIVE METHODS

6. SIGNAL DETECTION MEASURES

7. SCALING METHODS

8. MODEL COMPARISONS

Preface

The impetus for this book was a recurrent question posed not only by students but senior colleagues using psychophysics also for the first time: "Is there a book that explains how to do psychophysics?" The question suggested the need not only for a book that explained the theory behind psychophysical procedures, but that provided the practical tools for their implementation. In particular, there seemed to be a pressing need for a detailed and accessible exposition of how raw psychophysical responses are turned into meaningful measurements of sensory function.

The need for a practical book on psychophysics inevitably led to a second need: a comprehensive package of software for analyzing psychophysical data. The result was Palamedes. Although developed in conjunction with the book, Palamedes has since taken on a life of its own and is applicable to a much wider range of scenarios than are described in this book.

The first few chapters of the book are intended to introduce the basic concepts and terminology of psychophysics as well as familiarize readers with the full range of available psychophysical procedures. The remaining chapters focus on a series of specialist topics: psychometric functions, adaptive procedures, signal detection theory, scaling methods and statistical model comparisons. There is also a quick reference guide to the terms, concepts and many of the equations described in the book.

Writing this book has proved to be a considerable challenge for its authors. Much effort has been expended in the attempt to make accessible the theory behind different types of psychophysical data analysis. And because the meanings of many psychophysical terms and concepts are open to interpretation, we have on occasion had to improvise our own definitions (e.g. for the term 'appearance') and challenge existing conventions (e.g. by referring to a class of forced-choice tasks as 1AFC). Where we have challenged convention we have explained our reasoning, and hope that even if readers do not agree with our arguments, they will at the very least find them thought-provoking.

Acknowledgements

We are indebted to the following persons for kindly reviewing and providing insightful comments on earlier drafts of the individual chapters in the book: Curtis Baker, Athena Buckthought, Douglas Creelman, Kathleen Cullen, Mark Georgeson, Julie Harris, Mike Landy, Neil Macmillan, Tim Meese, Jonathan Peirce, Stéphane Rainville, Andrew Schofield, David Simmons, and Roger Watt.

Special thanks also to: Neil Macmillan and Douglas Creelman for helping one of the authors (FK) get to grips with the calculation of d' for same-different tasks (Chapter 6); Mark Georgeson for providing the derivation of the equation for the criterion measure $\ln\beta$ for a 2AFC task (also Chapter 6); Mark McCourt for providing the figures illustrating grating-induction (Chapter 3); Laurence Maloney for permission to develop and describe the routines for Maximum Likelihood Difference Scaling (Chapter 7).

About the Authors

Fred Kingdom is a Professor at McGill University conducting research into a variety of aspects of visual perception, including color vision, stereopsis, texture perception, contour-shape coding, the perception of transparency and visual illusions.

Nick Prins is an Associate Professor at the University of Mississippi specializing in visual texture perception, motion perception, contour-shape coding and the use of statistical methods in the collection and analysis of psychophysical data.

CHAPTER

1

Introduction and Aims

1.1 WHAT IS PSYCHOPHYSICS?

According to the online encyclopedia *Wikipedia*, psychophysics is "... a subdiscipline of psychology dealing with the relationship between physical stimuli and their subjective correlates, or percepts." The term psychophysics was first coined by Gustav Theodor Fechner (see front cover). In his *Elements of Psychophysics* (1860/1966) Fechner set out the principles of psychophysics, describing the various procedures that experimentalists use to map out the relationship between matter and mind. Although psychophysics is a methodology, it is also a research area in its own right, and a great deal of time is devoted to developing new psychophysical techniques and new methods for analyzing psychophysical data.

Psychophysics can be applied to any sensory system, whether vision, hearing, touch, taste or smell. This book primarily draws on research into the visual system to illustrate psychophysical principles, but for the most part the principles are applicable to all sensory domains.

1.2 AIMS OF THE BOOK

Broadly speaking, the book has three aims. The first is to provide newcomers to psychophysics with an overview of different psychophysical procedures in order to help them select the appropriate designs and analytical methods for their experiments. The second aim is to provide software for analyzing psychophysical data, and this is intended for both newcomers and experienced researchers alike. The third aim is to explain the theory behind the analyses, again mainly for newcomers, but also for experienced researchers who may be unfamiliar with a particular method or analysis. Thus, although *Psychophysics: A Practical Introduction* is primarily directed at newcomers to psychophysics, it is intended to provide sufficient new material, either in the form of software or theory, to engage the seasoned practitioner. To this end we have made every effort to make accessible, to both expert and non-expert alike, the theory behind a wide range of psychophysical procedures, analytical principles and mathematical computations: for example, Bayesian curve fitting; the calculation of d-primes (d'); maximum likelihood difference scaling; goodness-of-fit measurement; bootstrap analysis and likelihood-ratio testing, to name but a few. In short, the book is intended to be both practical and pedagogical.

The emphasis on practical implementation will hopefully offer the reader something not available in textbooks such as Gescheider's (1997) excellent *Psychophysics: The Fundamentals*. If there is a downside, however, it is that we do not delve as deeply into the relationship between psychophysical measurement and sensory function as *The Fundamentals* does, except when necessary to explain a particular psychophysical procedure. In this regard *A Practical Introduction* is not intended as a replacement for other textbooks on psychophysics, but as a complement to them, and readers are encouraged to read other relevant texts alongside our own.

Our approach of combining the practical and the pedagogical into a single book may not be to everyone's taste. Doubtless some would prefer to have the description of the software routines separate from the theory behind them. However we believe that by integrating the software with the theory, newcomers will be able to get a quick handle on the nuts-and-bolts of psychophysics methodology, the better to go on to grasp the underlying theory if and when they choose.

1.3 ORGANIZATION OF THE BOOK

The book can be roughly divided into two parts. Chapters 2 and 3 provide an overall framework and detailed breakdown of the variety of psychophysical procedures available to the researcher. Chapters 4–8 can be considered as technical chapters. They describe the software routines and background theory for five specialist topics: Psychometric Functions; Adaptive Methods; Signal Detection Measures; Scaling Methods; and Model Comparisons.

In Chapter 2 we provide an overview of some of the major varieties of psychophysical procedures, and offer a framework for classifying psychophysics experiments. The approach taken here is an unusual one. Psychophysical procedures are discussed in the context of a critical examination of the various dichotomies commonly used to differentiate psychophysics experiments: Class A versus Class B; objective versus subjective; Type 1 versus Type 2; performance versus appearance; forced-choice versus non-forced-choice; criterion-dependent versus criterion-free; detection versus discrimination; threshold versus suprathreshold. We consider whether any of these dichotomies could usefully form the basis of a fully-fledged classification scheme for psychophysics experiments, and conclude that one, the performance versus appearance distinction, is the best candidate.

Chapter 3 takes as its starting point the classification scheme outlined in Chapter 2, and expands on it by incorporating a further level of categorization based on the number of stimuli presented per trial. The expanded scheme serves as the framework for detailing a much wider range of psychophysical procedures than described in Chapter 2.

Four of the technical chapters, Chapters 4, 6, 7 and 8, are divided into two sections. Section A introduces the basic concepts of the topic and takes the reader through the Palamedes routines (see below) that perform the relevant data analyses. Section B provides more detail as well as the theory behind the analyses. The idea behind the Section A versus Section B distinction is that readers can learn about the basic concepts and their implementation without necessarily having to grasp the underlying theory, yet have the theory available to delve into if they want. For example, Section A of Chapter 4 describes how to fit psychometric functions and derive estimates of their critical parameters such as threshold and slope, while Section B describes the theory behind the various fitting procedures. Similarly, Section A in Chapter 6 outlines why d' measures are useful in psychophysics and how they can be calculated using Palamedes, while Section B describes the theory behind the calculations.

1.4 INTRODUCING PALAMEDES

According to Wikipedia, the Greek mythological figure Palamedes is said to have invented "… counting, currency, weights and measures, jokes, dice and a forerunner of chess called *pessoi*, as well as military ranks." The story goes that Palamedes also uncovered a ruse by Odysseus. Odysseus had promised Agamemnon that he would defend the marriage of Helen and Menelaus, but pretended to be insane to avoid having to honor his commitment. Unfortunately, Palamedes's unmasking of

Odysseus led to a gruesome end; he was stoned to death for being a traitor after Odysseus forged false evidence against him. We chose Palamedes as the name for the toolbox for his (presumed) contributions to the art of measurement, interest in stochastic processes (he did invent dice!), numerical skills, humor and wisdom. The Palamedes Swallowtail butterfly (*Papilio Palamedes*) on the front cover also provides the toolbox with an attractive icon.

Palamedes is a set of routines and demonstration programs written in MATLAB® for analyzing psychophysical data (Prins & Kingdom, 2009). The routines can be downloaded from www.palamedestoolbox.org. We recommend that you check the website periodically, because new and improved versions of the toolbox will be posted there for download. Chapters 4–8 explain how to use the routines and describe the theory behind them. The descriptions of Palamedes do not assume any knowledge of MATLAB, although a basic knowledge will certainly help. Moreover, Palamedes requires only basic MATLAB; the specialist toolboxes such as the Statistics Toolbox are not required. We have also tried to make the routines compatible with earlier versions of MATLAB, where necessary including alternative functions that are called when later versions are undetected.

It is important to bear in mind what Palamedes is *not*. It is not a package for generating stimuli, or for running experiments. In other words it is not a package for dealing with the "front-end" of a psychophysics experiment. The exceptions to this rule are the Palamedes routines for adaptive methods, which are designed to be incorporated into an actual experimental program, and the routines for generating stimulus lists for use in scaling experiments. But by-and-large, Palamedes is a different category of toolbox from the stimulus-generating toolboxes such as VideoToolbox (http://vision.nyu.edu/VideoToolbox/), PsychToolbox (http://psychtoolbox.org/wikka.php?wakka=HomePage), PsychoPy (http://www.psychopy.org, see also Peirce, 2007; 2009) and Psykinematix (http://psykinematix.kybervision.net/) (for a comprehensive list of such toolboxes see http://visionscience.com/documents/strasburger_files/strasburger.html). Although some of these toolboxes contain routines that perform similar functions to some of the routines in Palamedes, for example for fitting psychometric functions (PFs), they are in general complementary to, rather than in competition with Palamedes.

Of the few software packages that deal primarily with the analysis of psychophysical data, psignifit is perhaps the best known (http://bootstrap-software.org/psignifit/; see also Wichmann & Hill, 2001a,b). Like Palamedes, psignifit fits PFs, obtains errors for the parameter estimates of PFs using bootstrap analysis, and performs goodness-of-fit tests of PFs. Palamedes, however, does more; it has routines for calculating signal detection measures, implementing adaptive procedures and analyzing scaling data. For fitting PFs, some advantages of psignifit over Palamedes are that it executes much faster, its core function is a standalone executable file (i.e., it does not require MATLAB) and it has a few useful options that Palamedes does not have (notably, it allows one to perform a few different types of bootstrap, each with their own advantages). The advantage of Palamedes is that it can fit PFs to multiple conditions simultaneously, while providing the user considerable flexibility in defining a model to fit. Just to give one example, one might assume that the lapse rate and slope of the PF are equal between conditions, but that thresholds are not. Palamedes allows one to specify and implement such assumptions and fit the conditions accordingly. Palamedes can also be used to perform statistical comparisons between models. Examples are to test whether thresholds differ significantly between two or more conditions, to test whether it is reasonable to assume that slopes are equal between the conditions, to test whether the lapse rate differs significantly from zero (or any other specific value), etc. Finally, Palamedes allows one to test the goodness-of-fit of a user-defined model that describes performance across multiple conditions of an experiment.

1.4.1 Organization of Palamedes

All the Palamedes routines are prefixed by an identifier **PAL**, to avoid confusion with the routines used by MATLAB. After **PAL**, many routine names contain an acronym for the class of procedure they implement. Table 1.1 lists the acronyms currently in the toolbox, what they stand for, and the book chapter where they are described. In addition to the routines with specialist acronyms, there are a number of general-purpose routines.

TABLE 1.1 Acronyms used in Palamedes, their meaning and the chapter in which they are described

Acronym	Meaning	Chapter
PF	Psychometric function	4
PFBA	Psychometric function: Bayesian	4
PFML	Psychometric function: maximum likelihood	4, 8
AMPM	Adaptive methods: psi method	5
AMRF	Adaptive methods: running fit	5
AMUD	Adaptive methods: up/down	5
SDT	Signal detection theory	6
MLDS	Maximum likelihood difference scaling	7
PFLR	Psychometric function: likelihood ratio	8

1.4.2 Functions and Demonstration Programs in Palamedes

1.4.2.1 *Functions*

In MATLAB there is a distinction between a function and a script. A function accepts one or more input arguments, performs a set of operations and returns one or more output arguments. Typically, Palamedes functions are called as follows:

```
>>[x y z] = PAL_FunctionName(a,b,c);
```

where **a**, **b** and **c** are the input arguments, and **x**, **y** and **z** the output arguments. In general, the input arguments are "arrays." Arrays are simply listings of numbers. A scalar is a single number, e.g., 10, 1.5, 1.0e–15. A vector is a one-dimensional array of numbers. A matrix is a two-dimensional array of numbers. It will help you to think of all as being arrays. As a matter of fact, MATLAB represents all as two-dimensional arrays. That is, a scalar is represented as a 1×1 (1 row $\times$ 1 column) array, vectors either as an $m \times 1$ array or a $1 \times n$ array, and a matrix as an $m \times n$ array. Arrays can also have more than two dimensions.

In order to demonstrate the general usage of functions in MATLAB, Palamedes includes a function named **PAL_ExampleFunction** which takes two arrays of any dimensionality as input arguments and returns the sum, the difference, the product, and the ratio of the numbers in the arrays corresponding to the input arguments. For any function in Palamedes you can get some information as to its usage by typing **help** followed by the name of the function:

```
>>help PAL_ExampleFunction
```

MATLAB returns:

```
PAL_ExampleFunction calculates the sum, difference, product,
and ratio of two scalars, vectors or matrices.

syntax: [sum difference product ratio] = ...
PAL_ExampleFunction(array1, array2)

This function serves no purpose other than to demonstrate the
general usage of Matlab functions.
```

For example, if we type and execute:

```
[sum difference product ratio] = PAL_ExampleFunction(10, 5);
```

MATLAB will assign the arithmetic sum of the input arguments to a variable labeled **sum**, the difference to **difference**, etc. In case the variable **sum** did not previously exist, it will have been created when the function was called. In case it did exist, its previous value will be overwritten (and thus lost). We can inquire about the value of a variable by typing its name, followed by <return>:

```
>>sum
```

MATLAB returns:

```
sum = 15
```

We can use any name for the returned arguments. For example, typing:

```
>>[s d p r] = PAL_ExampleFunction(10,5)
```

creates a variable **s** to store the sum, etc.

Instead of passing values directly to the function, we can assign the values to variables and pass the name of the variables instead. For example the series of commands:

```
>>a = 10;
>>b = 5;
>>[sum difference product ratio] = PAL_ExampleFunction(a, b);
```

generates the same result as before. You can also assign a single alphanumeric name to vectors and matrices. For example, to create a vector called **vect1** with values 1, −2, 4, and 105 one can simply type and follow with a <return>:

```
>> vect1 = [1 -2 4 105]
```

Note the square, not round brackets. **vect1** can then be entered as an argument to a routine, provided the routine is set up to accept a 1×4 vector. To create a matrix called **matrix1** containing two columns and three rows of numbers, type and follow with a <return>, for example:

```
>> matrix1 = [.01 .02; .04 .05; 0.06 0.09]
```

where the semicolon separates the values for different rows. Again, **matrix1** can now be entered as an argument, provided the routine accepts a 3×2 (rows by columns) matrix.

Whenever a function returns more than one argument, we do not need to assign them all to a variable. Let's say we are interested in the sum and the difference of two matrices only. We can type:

```
>>[sum difference] = PAL_ExampleFunction([1 2; 3 4], [5 6; ...
7 8]);
```

1.4.2.2 Demonstration Programs

A separate set of Palamedes routines are suffixed by **_Demo**. These are demonstration scripts that in general combine a number of Palamedes function routines into a sequence to demonstrate some aspect of their combined operation. They produce a variety of types of output to the screen, such as numbers with headings, graphs, etc. While these programs do not take arguments when they are called, the user might be prompted to enter something when the program is run, e.g.:

```
>>PAL_Example_Demo
Enter a vector of stimulus levels
```

Then the user might enter something like `[.1 .2 .3]`. After pressing return there will be some form of output, for example data with headings, a graph, or both.

1.4.3 Error Messages in Palamedes

The Palamedes toolbox is not particularly resistant to user error. Incorrect usage will more often result in a termination of execution accompanied by an abstract error message than it will in a gentle warning or a suggestion for proper usage. As an example, let us pass some inappropriate arguments to our example function and see what happens. We will pass two arrays to it of unequal size:

```
>>a = [1 2 3];
>>b = [4 5];
>>sum = PAL_ExampleFunction(a, b);
```

MATLAB returns:

```
??? Error using ==> unknown
Matrix dimensions must agree.
Error in ==> PAL_ExampleFunction at 15
sum = array1 + array2;
```

This is actually an error message generated by a resident MATLAB function, not a Palamedes function. Palamedes routines rely on many resident MATLAB functions and operators (such as "+"), and error messages you see will typically be generated by these resident MATLAB routines. In this case, the problem arose when **PAL_ExampleFunction** attempted to use the "+"operator of MATLAB to add two arrays that are not of equal size.

References

Fechner, G. (1860/1966). *Elements of Psychophysics*. Hilt, Rinehart & Winston, Inc.

Gescheider, G. A. (1997). Psychophysics: The Fundamentals. Lawrence Erlbaum Associates, Mahwah, New Jersey.

Peirce, J. W. (2007). PsychoPy – Psychophysics software in Python. *Journal of Neuroscience Methods, 162*(1–2), 8–13.

Peirce, J. W. (2009). Generating stimuli for neuroscience using PsychoPy doi:10.3389/neuro.11.010.2008. *Frontiers in Neuroinformatics*, 2, 10.

Prins, N., & Kingdom, F.A.A. (2009). Palamedes: MATLAB routines for analyzing psychophysical data. http://www.palamedestoolbox.org.

Wichmann, F. A., & Hill, N. J. (2001a). The psychometric function: I. Fitting, sampling and goodness-of-fit. *Perception and Psychophysics, 63*, 1293–1313.

Wichmann, F. A., & Hill, N. J. (2001b). The psychometric function: II. Bootstrap-based confidence intervals and sampling. *Perception and Psychophysics, 63*, 1314–1329.

CHAPTER

2

Classifying Psychophysical Experiments

2.1 INTRODUCTION

This chapter introduces some of the main classes of psychophysical procedure and proposes a scheme for classifying psychophysical experiments. The aim is not so much to judge the pros and cons of different psychophysical procedures – this will form part of the subject matter of the next chapter – but to consider how different psychophysical procedures fit together into the jigsaw we call psychophysics. The resulting classification scheme is arrived at through a critical examination of the familiar "dichotomies" that make up the vernacular of psychophysics, for example "Class A" versus "Class B" observations, "Type 1" versus "Type 2" tasks, "forced-choice" versus "non-forced-choice," etc. These dichotomies do not in every case mean the same thing to all people, so the aim here is both to clarify what they mean and to decide which of them, if any, might be useful as categories in a classification scheme.

Why a classification scheme? After all, the seasoned practitioner designs a psychophysics experiment based on implicit knowledge accumulated over years of research experience as to what is available, what is appropriate, and what is valid, given the question about visual function being asked. And that is how it should be. However, a framework that captures both the critical differences and intimate relationships between different experimental approaches should be useful for newcomers to the field, helping them to select the appropriate design from what might first seem a bewildering array of possibilities. Thinking about a classification scheme is also a useful intellectual exercise, not only for those of us who like to categorize things, put them into boxes and attach labels to them, but for anyone interested in gaining a deeper understanding of psychophysics. But before discussing the dichotomies, let us first consider the components that make up a psychophysics experiment.

2.2 TASKS, METHODS, AND MEASURES

Although the measurement at the end of a psychophysics experiment reflects more than anything else the particular question about sensory function being asked, other components of the experiment, in particular the stimulus and the observer's task, must be carefully tailored to achieve the experimental goal. A psychophysics experiment consists of a number of components, and we have opted for the following breakdown: stimulus; task; method; analysis; and measure (Figure 2.1). To illustrate our use of these terms, consider one of the most basic experiments in the study of vision: the measurement of a "contrast detection threshold." A contrast detection threshold is defined as the minimum amount of contrast necessary for a stimulus to be just detectable. Figure 2.2 illustrates the idea for a stimulus consisting of a patch on a uniform background. The precise form of the stimulus must, of course, be tailored to the specific question about sensory function being asked, so

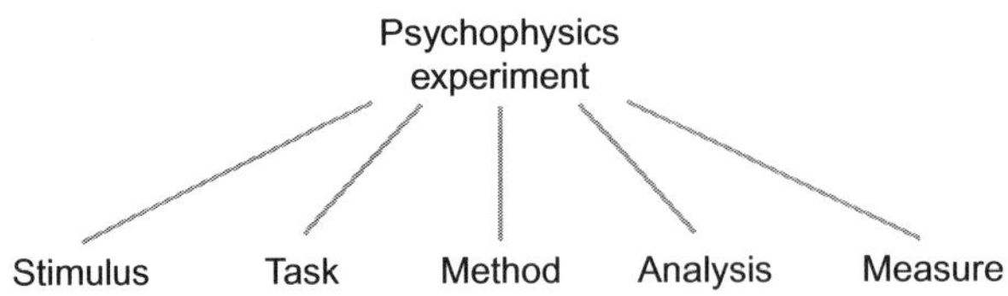

FIGURE 2.1 Components of a psychophysics experiment.

let us simply assume that the patch is the appropriate stimulus. The contrast of the patch can be measured in terms of Weber contrast, defined as the difference between the luminance of the patch and its background, ΔL, divided by the luminance of the background L_b, i.e. $\Delta L/L_b$. Thus the contrast detection threshold is the smallest value of Weber contrast an observer requires to detect the patch. There are many procedures for measuring a contrast detection threshold, and each involves a different task for the observer. Before the advent of digital computers, a common method was to display the stimulus on an oscilloscope screen and require observers to adjust the contrast with a dial until the stimulus was just visible against the background. The just-visible contrast would be recorded as the measure of the contrast detection threshold. For this type of procedure, the task and the method are collectively termed the "method of adjustment."

Nowadays the preferred approach is to present the stimulus on a computer display and use a standard "two-interval forced-choice," or 2IFC task. Using this procedure, two stimuli are presented briefly on each trial, one a blank screen, the other the test patch. The order of stimulus presentation – blank screen followed by test patch or test patch followed by blank screen – is unknown to the observer (although of course 'known' to the computer), and is typically random or quasi-random. The two stimuli are presented consecutively, and the observer must choose the interval containing the test patch, indicating his or her choice by pressing a key. The computer keeps a record of the contrast of the patch for each trial, along with the observer's

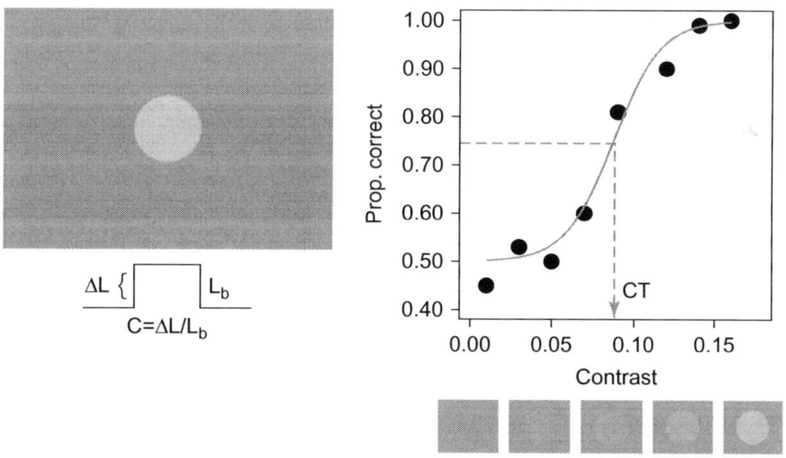

FIGURE 2.2 Top: circular test patch on a uniform background. Bottom: luminance profile of patch and the definition of Weber contrast. Right: results of a standard two-interval-forced-choice (2IFC) experiment. The various stimulus contrasts are illustrated on the abscissa. Black circles give the proportion of correct responses for each contrast. The green line is the best fit of a psychometric function, and the calculated contrast detection threshold (CT) is indicated by the arrow. See text for further details. L = luminance; L_b = luminance of background; ΔL = difference in luminance between patch and background; C = Weber contrast.

response, which is scored as either "correct" or "incorrect." A given experimental session might consist of, say, 100 trials, and a number of different patch contrasts would be presented in random order.

With the standard 2IFC task, different methods are available for selecting the contrasts presented on each trial. On the one hand, they can be preselected before the experiment; for example, ten contrasts ranging from 0.01 to 0.1 at 0.01 intervals. If preselected, the ten stimuli at each contrast would be presented in random order during the session, making 100 trials in total. This is known as the "method of constants." At the end of each session the computer calculates the number of correct responses for each contrast. Typically, there would be a number of sessions and the overall proportion correct across sessions for each patch contrast calculated, then plotted on a graph as shown for the hypothetical data in Figure 2.2. On the other hand, one can use an "adaptive" (or "staircase") method, in which the contrast selected on each trial is determined by the observer's responses on previous trials. The idea is that the computer "homes in" on the contrasts that are close to the observer's contrast detection threshold, thus not wasting too many trials presenting stimuli that are either too easy or too hard to see. Adaptive methods are the subject of Chapter 5.

The term "analysis" refers to how the data collected during an experiment are converted into measurements. For example, with the method of adjustment the observer's settings might be averaged to obtain the threshold. On the other hand, using the 2IFC procedure in conjunction with the method of constants, the proportion correct data may be fitted with a function whose shape is chosen to approximately match the data. The fitting procedure estimates the contrast resulting in a criterion proportion correct, such as 0.75 or 75%, and this is the estimate of the contrast detection threshold, as shown for our hypothetical data in Figure 2.2. Procedures for fitting psychometric functions are discussed in Chapter 4.

To summarize: using the example of an experiment aimed at measuring a contrast detection threshold for a patch on a uniform background, the components of a psychophysical experiment are as follows. The "stimulus" is a uniform patch of given spatial dimensions and of various contrasts. Example "tasks" include adjustment and 2IFC. For the adjustment task, the "method" is the method of adjustment, while for the 2IFC task the methods include the method of constants and adaptive methods. In the case of the method of adjustment, the "analysis" might consist of averaging a set of adjustments, whereas for the 2IFC task it might consist of fitting a psychometric function to the proportion correct responses as a function of contrast. For the 2IFC task in conjunction with an adaptive method, the analysis might involve averaging contrasts, or it might involve fitting a psychometric function. The "measure" in all cases is a contrast detection threshold, although other measures may also be extracted, such as an estimate of the variability or "error" on the threshold and the slope of the psychometric function.

The term "procedure" is used ubiquitously in psychophysics, and can refer variously to the task, method, analysis or some combination of these. Similarly the term "method" has broad usage. The other terms in our component breakdown are also often used interchangeably. For example, the task in the contrast detection threshold experiment, which we termed adjustment or 2IFC, is sometimes termed a "detection" task and sometimes a "threshold" task, even though in our taxonomy these terms refer to the measure. The lesson here is that one must be prepared to be flexible in the use of psychophysics terminology, and not overly constrained by any predefined scheme.

Our next step is to consider some of the common dichotomies used to differentiate psychophysical experiments. The aim here is to introduce some familiar terms used in psychophysics, illustrate other classes of psychophysical experiment besides contrast detection and examine which, if any, of the dichotomies might be candidates for a classification scheme for psychophysical experiments.

2.3 DICHOTOMIES

2.3.1 "Class A" Versus "Class B" Observations

An influential dichotomy introduced some years ago by Brindley (1970) is that between "Class A" and "Class B" psychophysical observations. Although one rarely hears these terms today, they are important to our understanding of the relationship between psychophysical measurement and sensory function. Brindley used the term "observation" to describe the perceptual state of the observer while executing a psychophysical task. The distinction between Class A and Class B attempted to identify how directly a psychophysical observation related to the underlying mental processes involved. Brindley framed the distinction in terms of a comparison of sensations: a Class A observation refers to the situation in which two physically different stimuli are perceptually indistinguishable; whereas a Class B observation refers to all other situations.

The best way to understand the distinction between Class A and Class B is by an example, and we have adopted Gescheider's (1997) example of the Rayleigh match (Rayleigh, 1881; Thomas & Mollon, 2004). Rayleigh matches are used to identify and study certain types of color vision deficiency (e.g., Shevell, Sun & Neitz, 2008) although for the present discussion the purpose of a Rayleigh match is less important than the nature of the measurement itself. Figure 2.3 shows a bipartite circular stimulus, one half consisting of a mixture of red and green monochromatic lights, the other half a yellow monochromatic light.[1] The observer has free reign to adjust

[1]Because the lights are monochromatic, i.e., narrow band in wavelength, this experiment cannot be conducted on a CRT monitor, because CRT phosphors are relatively broadband in wavelength. Instead an apparatus is required that can produce monochromatic lights, such as a Nagel Anomaloscope or a Maxwellian view system.

both the intensity of the yellow light, as well as the relative intensities of the red and green lights. The task is to adjust the intensities of the lights until the two halves of the stimulus appear identical, as illustrated in the top of the figure. In color vision, two stimuli with different spectral (i.e., wavelength) compositions that nevertheless look identical are termed "metamers." According to Brindley, metameric matches, such as the Rayleigh match, are Class A observations. The identification of an observation as Class A accords with the idea that when two stimuli look identical to the eye they elicit identical neural responses in the brain. Since the neural responses are

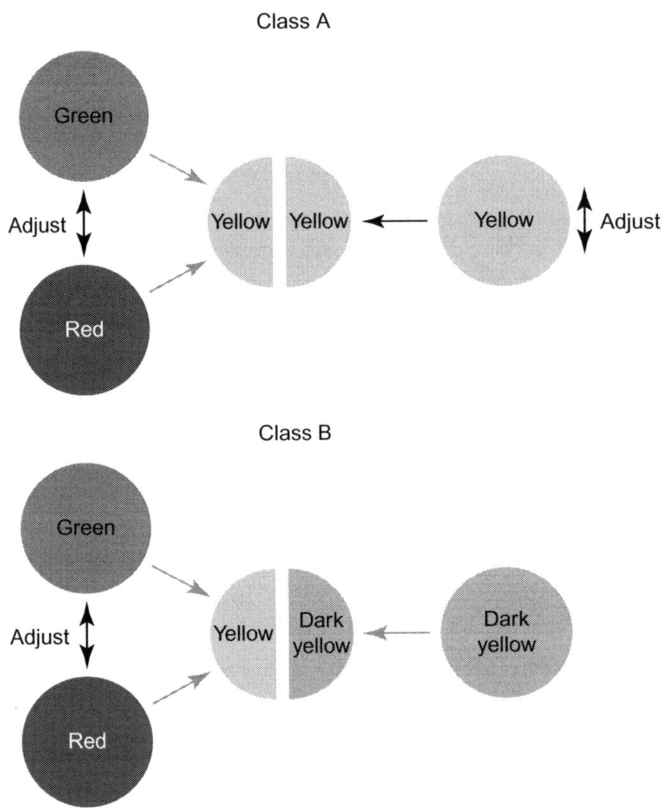

FIGURE 2.3 The Rayleigh match illustrates the difference between a Class A and Class B psychophysical observation. For Class A, the observer adjusts both the intensity of the yellow light in the right half of the bipartite field, as well as the relative intensities of the red and green mixture in the left half of the bipartite field, until the two halves appear identical. For Class B, the observer adjusts only the relative intensities of the red and green lights to match the hue of a yellow light that is different in brightness. See separate color plate section for the full color version of this figure.

identical, Brindley argues, it is relatively straightforward to map the physical characteristics of the stimuli onto their internal neural representations.

An example of a Class B observation is shown at the bottom of Figure 2.3. This time the observer has no control over the intensity of the yellow light – only control over the relative intensities of the red and green lights. The task is to match the hue (or perceived chromaticity) of the two halves of the stimulus, but under the constraint that the intensity (or brightness) of the two halves are different. Thus, the two halves will never look identical and, therefore, according to Brindley, neither will the neural responses they elicit. Brindley was keen to point out that one must not conclude that Class B observations are inferior to Class A observations. Our example Class B observation is not a necessary evil given defective equipment! On the contrary, we may want to know which spectral combinations match in hue when their brightnesses are very different, precisely in order to understand the way that hue and brightness interact. In any case, the aim here is not to judge the relative merits of Class A and Class B observations (for a discussion of this see Brindley, 1970), but to illustrate what the terms mean.

What other types of psychophysical experiment are Class A and Class B? According to Brindley, experiments that measure thresholds, such as the contrast detection threshold experiment discussed above, are Class A. That a threshold is Class A might not be intuitively obvious. The argument goes something like this. There are two states: stimulus present and stimulus absent. As the stimulus contrast is decreased to a point where it is below threshold, the observation passes from one in which the two states are discriminable, to one in which they are indiscriminable. The fact that the two states may be indiscriminable even though physically different (the stimulus is still present even though below threshold) makes the observation Class A. Two other examples of Class A observations that obey the same principle are shown in Figure 2.4.

Class B observations characterize many types of psychophysical procedure. Following our example Class B observation in Figure 2.3, any experiment that involves matching two stimuli that remain perceptibly different on completion of the match is Class B. Consider, for example, the brightness-matching experiment illustrated in Figure 2.5. The aim of the experiment is to understand how the brightness or perceived luminance of a circular test patch is influenced by the luminance of its surround. As a rule, increasing the luminance of a surround annulus causes the region it encloses to decrease in brightness, i.e., become dimmer. One way to measure the amount of dimming is to adjust the luminance (and since the background luminance is fixed, also the contrast) of a second patch until it appears equal in brightness to the test patch. The second patch can be thought of as a psychophysical "ruler." When the match is set equal in brightness to the test, the patches are said to be at the "point of subjective equality," or PSE. The luminances of the test and match patches at the PSE will not necessarily be the same; indeed it is precisely that they will be, in most instances, different that is of interest. The difference in luminance

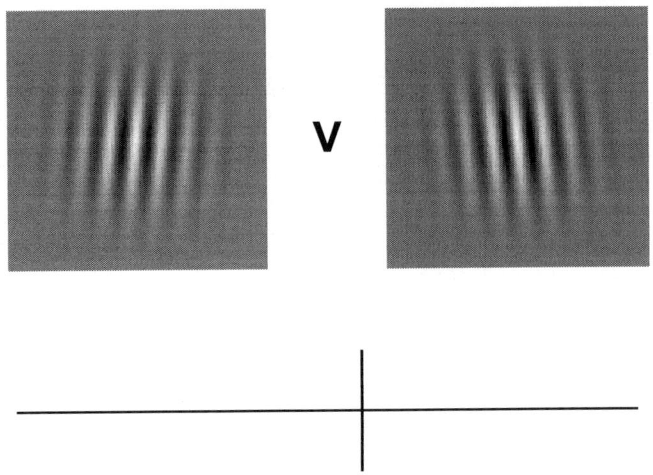

FIGURE 2.4 Two other examples of Class A observations. Top: orientation discrimination task. The observer is required to discriminate between two gratings that differ in orientation, and a threshold orientation difference is measured. Bottom: line bisection task. The observer is required to position the vertical line midway along the horizontal line. The precision or variability in the observer's settings is a measure of their line-bisection acuity.

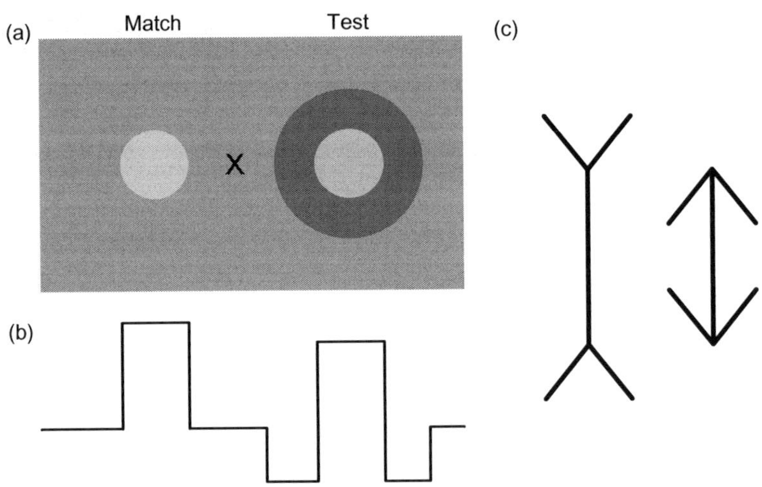

FIGURE 2.5 Two examples of Class B observations. In (a) the goal of the experiment is to find the point of subjective equality (PSE) in brightness between the fixed test and variable match patches, as a function of the luminance (and hence contrast) of the surround annulus; (b) shows the approximate luminance profile of the stimulus; (c) Muller–Lyer illusion. The two center lines are physically identical, but appear different in length. The experiment described in the text measures the relative lengths of the two center lines at which they appear equal in length.

between the test and match at the PSE tells us something about the effect of context on brightness, the "context" in this example being the annulus. This type of experiment is sometimes referred to as "asymmetric brightness matching," because the test and match patches are set in different contexts (e.g., Blakeslee & McCourt, 1997; Hong & Shevell, 2004).

It might be tempting to think of an asymmetric brightness match as a Class A observation, owing to the fact that it differs in one important respect from the Class B observation in the brightness-unmatched version of the Rayleigh match discussed above. In the brightness-unmatched version of the Rayleigh match, the stimulus region that is matched in hue is *also* the region that is held different in brightness. In an asymmetric brightness-matching experiment on the other hand, the stimulus region that is matched in brightness is *not* the region that defines the difference between the test and match; this is the annulus. However, one cannot "ignore" the annulus when deciding whether the observation is Class A or Class B simply because it is not the part of the stimulus to which the observation is directed. Asymmetric brightness matches are Class B observations because, even when the stimuli are matched, they are recognizably different owing to the fact that one has an annulus and the other has not.

Another example of a Class B observation is the Muller–Lyer illusion shown in Figure 2.5c, a geometric visual illusion that has received much attention (e.g., Morgan, Hole, & Glennerster, 1990). The lengths of the center lines in the two figures are the same, yet they appear different due to the arrangement of fins at the ends. One method for measuring the size of the illusion is to require observers to adjust the length of one of the center lines until it matches the perceived length of the other. The physical difference in length at the PSE, which could be expressed as a raw, proportional or percentage difference, measures the size of the illusion. The misperception of relative line length in the Muller–Lyer figures is a Class B observation, because even when the lines are adjusted to make them perceptually equal in length the figures remain recognizably different, owing to their different fin arrangements.

One further example of a Class B observation is magnitude estimation – the procedure whereby observers provide a numerical estimate of the perceived magnitude of a stimulus along some dimension, e.g., contrast, speed, depth, size, etc. Magnitude estimation is Class B, because the perception of the stimulus and the judgement of its magnitude involve different mental modalities.

An interesting case that at first defies classification into Class A or Class B is illustrated in Figure 2.6. The observer's task is to discriminate the mean orientation of random arrays of line elements, whose mean orientations are right- and left-of-vertical (e.g., Dakin, 2001). Below threshold, the mean orientations of the two arrays are by definition indiscriminable, yet the two arrays are still perceptibly different in terms of their element arrangements. In the previously-mentioned Class B examples the "other" dimension – brightness in the case of the Rayleigh match, annulus luminance in the case of the brightness-matching experiment – was relevant to the task. However, in the mean-orientation-discrimination experiment the "other"

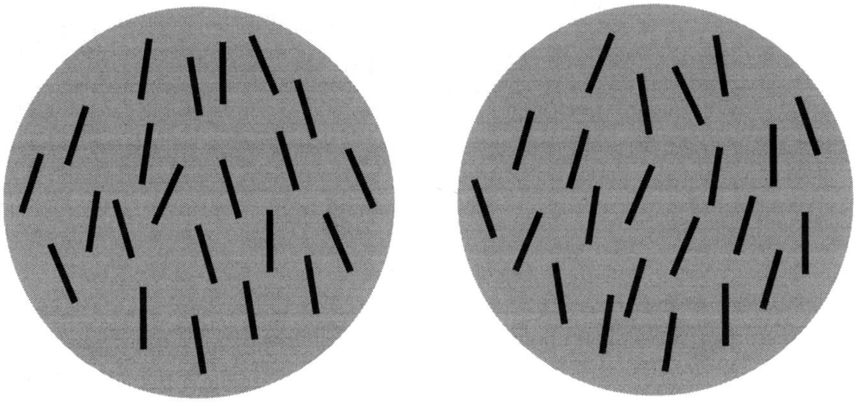

FIGURE 2.6 Class A or Class B? The observer's task is to decide which stimulus contains elements that are on average left-oblique. When the difference in mean element orientation is below threshold, the stimuli are identical in terms of their perceived mean orientation, yet are discriminable on the basis of the arrangements of elements.

dimension – element position – is irrelevant. Does the fact that element arrangement is irrelevant make it Class A, or does the fact that the stimuli are discriminable below threshold on the basis of element arrangement make it Class B? Readers can make up their own mind.

In conclusion, the Class A versus Class B distinction is important for understanding the relationship between psychophysical measurement and sensory function. However, we have chosen not to use it as a basis for classifying psychophysics experiments, partly because there are cases that seem to us hard to classify into Class A or Class B, and partly because other dichotomies better capture for us the critical differences between psychophysical experiments.

2.3.2 "Objective" Versus "Subjective"

Although rarely used in journal articles, the terms objective and subjective are common parlance among psychophysicists, so it is worth examining what they mean. The terms tend to be value-laden, with the connotation that objective is "good" and subjective "bad." Whether or not this is intended, the objective versus subjective dichotomy is inherently problematic when applied to psychophysics. All psychophysical experiments are in a trivial sense subjective, because they measure what is going on inside the head, and if this is the intended meaning of the term then the distinction is redundant. The dichotomy is more often invoked, however, to differentiate between different types of psychophysical procedure. The distinction has been used variously to characterize Class A versus Class B observations, tasks for which there is versus tasks

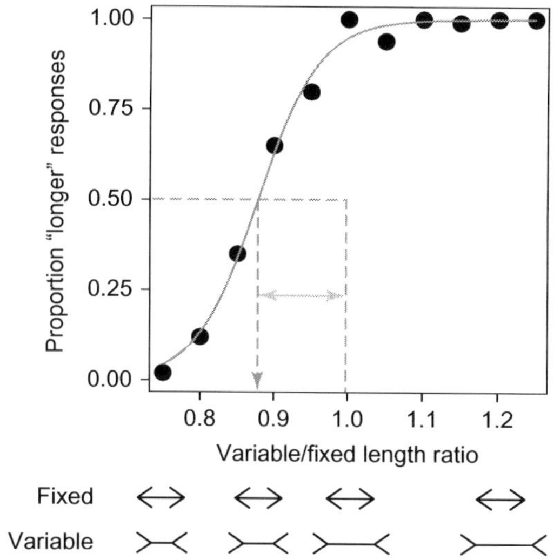

FIGURE 2.7 Results of a hypothetical experiment aimed at measuring the size of the Muller–Lyer illusion using a forced-choice procedure and the method of constant stimuli. The critical measurement is the PSE, or point of subjective equality between the lengths of the center lines in the fixed test and variable comparison stimuli. The graph plots the proportion of times subjects perceive the variable stimulus as "longer." The continuous line through the data is the best-fitting logistic function (see Chapter 4). The value of 1.0 on the abscissa indicates the point where the fixed and variable lengths are physically equal. The PSE is calculated as the variable length at which the fixed and variable lengths appear equal, indicated by the vertical green arrow. The horizontal green-arrowed line is a measure of the size of the illusion.

for which there is not a correct and an incorrect response, forced-choice versus non-forced-choice procedures, and criterion-dependent versus criterion-free procedures. We have already discussed the Class A versus Class B distinction, so in what follows we primarily concentrate on the other usages of the objective-subjective distinction.

Consider how the objective–subjective distinction might apply to the Muller–Lyer illusion mentioned above. As with the contrast detection threshold experiment, there is more than one way to measure the size of the illusion. The adjustment method is one way. A forced-choice procedure is another. Using forced-choice, both stimuli are presented as alternatives during the trial. On each trial the length of the center line of one stimulus is selected from a pre-specified set (call this the variable stimulus), while the length of the center line of the other stimulus is fixed (call this the fixed stimulus). The observer's task is to indicate the stimulus perceived to have the longer center line. Figure 2.7 shows hypothetical results from such an experiment. Each data point represents the proportion of times the variable stimulus is perceived as longer, as a function of its center line length relative to that of the fixed stimulus. At a relative length of 1, meaning that the center lines are *physically* the same, the observer perceives the variable stimulus as longer almost 100% of the time. However, at a relative length of about 0.88 the observer chooses the variable stimulus as longer only 50% of the time. Thus, the PSE is 0.88.

Even though this example illustrates how the Muller–Lyer illusion, like the contrast threshold experiment, can be measured using a forced-choice procedure, there is an important difference between the two experiments. Whereas in the forced-choice contrast detection threshold experiment there is a correct and an incorrect response on every trial, there is no correct or incorrect response for the Muller–Lyer trials. Whatever response the observer makes on a Muller–Lyer trial, it is meaningless to score it as correct or incorrect, at least given the goal of the experiment which is to measure a PSE. Observers unused to doing psychophysics often have difficulty grasping this idea, and even when told repeatedly that there is no correct and incorrect answer, insist on asking at the end of the experiment how many trials they scored correct!

For some researchers, judgements that cannot be evaluated in terms of being correct and incorrect are more subjective (or less objective) than those that can. This view presumably arises because judgements in correct-response experiments are evaluated against an "external" benchmark; the stimulus on each trial really *is* present or absent, or really *is* left or right oblique. The benchmark for tasks where there is no correct and incorrect response on the other hand is purely "internal;" the line only *appears* to be longer, or the patch only *appears* to be brighter. Psychophysical tasks that have correct and incorrect responses are termed Type 1, and those that do not are termed Type 2, a dichotomy to which we shall return in the next section.

For other researchers, however, the objective–subjective distinction is more to do with the method of data collection than with the nature of the measurement. Some researchers feel that forced-choice methods are inherently more objective than non-forced-choice methods, irrespective of whether they are Type 1 or Type 2. According to this point of view, both the contrast detection threshold and Muller–Lyer illusion experiments are more objective when using a forced-choice compared to an adjustment procedure.

Why might forced-choice experiments be considered more objective than non-forced-choice experiments? It could be argued that forced-choice methods provide more "accurate" estimates of thresholds and PSEs than those obtained from non-forced-choice methods. Accuracy, in this context, refers to how close the measure is to its "true" value. There is a problem, however, with this argument. How does one determine whether one method is more accurate than another? This is not easy to answer, particularly for PSEs. Another reason why forced-choice methods might be considered more objective is that they are more "precise." Precision refers to the variability in the measurement. With the method of adjustment, precision is typically calculated from the variance, or more usually the standard deviation of the observer's settings, with a small standard deviation indicating a high precision. With forced-choice methods, precision is typically measured by the steepness or slope of the psychometric function (see Figure 2.7). The slope of the psychometric function is inversely proportional to the standard deviation parameter in the function used to fit the data, so in this case a small standard deviation is an indication of high precision (see Chapter 4 for details). In principle, therefore, one could compare the precisions of adjustment and forced-choice procedures, and from the result argue that one is more objective than the other. However, even this is problematic. Suppose, for example, that the forced-choice procedure proved to be the more precise, but the experiment took much longer. One could argue that the superior precision was due to the longer experimental time, not the difference in method *per se*.

All of the above arguments lead us to conclude that the distinction between objective and subjective is too loosely-defined and inherently problematic to use as a basis for classifying psychophysical experiments.

2.3.3 "Type 1" Versus "Type 2"

In the previous section, we drew attention to another important distinction, that between experiments for which there is and experiments for which there is not a correct and an incorrect response on each trial. This distinction has been termed Type 1 versus Type 2 (Sperling, 2008; see also Sperling, Dosher & Landy, 1990).[2] The forced-choice version of the contrast threshold experiment described above is therefore Type 1, whereas the brightness-matching and Muller–Lyer experiments are Type 2. The term Type 2 has also been used to refer to observers, own judgements of their Type 1 decisions (Galvin et al., 2003). In this case, the Type 2 judgement might be a rating of, say, 1–5, or a binary judgement such as "confident" or "not confident," in reference to the decision made in a correct-answer forced-choice task.

As discussed in the previous section, the Type 1 versus Type 2 distinction may, for some researchers, be synonymous with objective versus subjective. However, Type 1 and Type 2 are not synonymous with Class A and Class B. The Rayleigh match experiment described above is Class A, but is Type 2 because there is no "correct" match.

The Type 1 versus Type 2 dichotomy is an important one in psychophysics. It dictates, for example, whether observers can be provided with feedback during an experiment, such as a tone for a correct but not an incorrect response. However, one should not conclude that Type 1 is "better;" the value of Rayleigh matches for understanding color deficiency is a clear case in point. Moreover, as we argue in the next section, Type 1 versus Type 2 is not for us the most important distinction in psychophysics.

[2]Note that the dichotomy is not the same as the Type I versus Type II error dichotomy used in statistical inference testing.

2.3.4 "Performance" Versus "Appearance"

A distinction closely related, but not synonymous with, Type 1 versus Type 2 is "performance" versus "appearance." Performance-based tasks measure "aptitude," i.e., "how good" an observer is at a particular task. For example, suppose one measures contrast detection thresholds for two sizes of patch, call them "small" and "big." If thresholds for the big patch are found to be lower than those for the small patch, one can conclude that observers are better at detecting big patches than small ones. By the same token, if orientation discrimination thresholds are found to be lower in central than in peripheral vision, one can conclude that orientation discrimination is better in central vision than in the periphery. In these examples, performance measures the "limits" of our perception. On the other hand, suppose we measure the Muller–Lyer illusion for two different fin angles, say 45 and 60 degrees relative to the center line, and find that the illusion is bigger for the 45 degree fins. It would be meaningless to conclude that we are "better" at the Muller–Lyer with the 45 degree fins. PSEs are not aptitudes. For this reason the Muller–Lyer experiment is best considered as measuring stimulus "appearance." A simple heuristic can be used to decide whether a psychophysical experiment/task/measurement is performance-based or appearance-based. If the measurement can be meaningfully considered to be better under one condition than under another, then it is a performance measure, if not it is an appearance measure. This still leaves open the question of a precise definition of appearance, other than "not performance." Appearance is not however an easy term to define, but in many instances one can usefully think of appearance as measuring the "apparent" magnitude of some stimulus dimension.

Sometimes the same psychophysical task can be used to measure both performance *and* appearance. Consider the vernier alignment task illustrated in Figure 2.8 applied to two stimulus arrangements, A and B. On each trial the observer decides whether the upper black line lies to the left (or to the right) of the lower black line. The goal of experiment A is to measure vernier "acuity," defined as the smallest misalignment that can be detected. The goal of experiment B, on the other hand, is to measure the effect of the flanking white lines on the relative perceived position of the black lines. The white lines in B have a repulsive effect, causing the black lines to appear slightly shifted from their normal perceived position in a direction away from that of the white lines (e.g., Badcock & Westheimer, 1984). Hypothetical data for both tasks are shown in the graph on the right, and are fitted with logistic functions (see Chapter 4).

For experiment A, vernier acuity can be calculated as the horizontal line separation which results in a proportion of 0.75 "left" responses, the point on the graph's abscissa indicated by the green arrow, and usually termed the vernier "threshold." Sometimes, however, the value shown by the green arrow is not a good estimate of the observer's vernier threshold. The observer will sometimes have a small bias

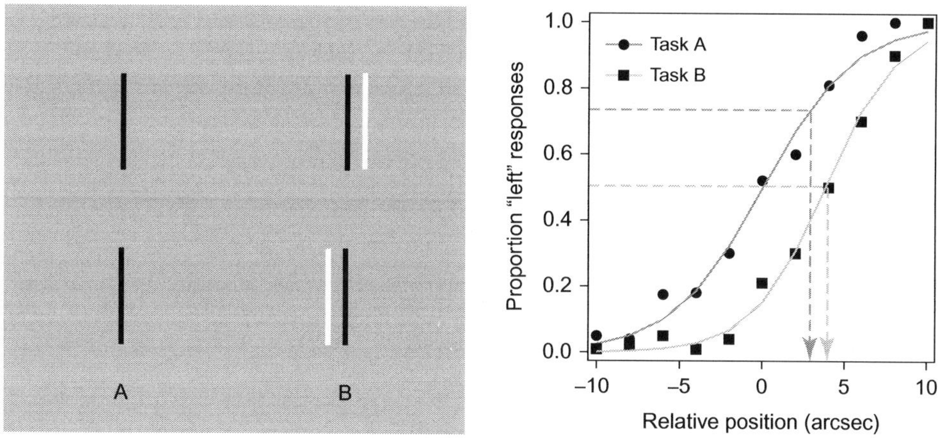

FIGURE 2.8 Left: stimulus arrangements for two vernier alignment experiments. Right: hypothetical data from each experiment. The abscissa plots the horizontal physical separation between the black lines, with positive values indicating that the top line is physically to the left of the bottom line and negative values that the top line is physically to the right. The ordinate gives the proportion of times the observer responds that the top line is "left." The continuous curves are best-fitting logistic functions. The green arrows indicate for Task A the vernier threshold and for Task B the point-of-subjective alignment.

towards perceiving the two lines as aligned when they are in fact slightly misaligned. In other words, the "point of subjective alignment" or PSA may not be zero, but a small positive or negative value. A non-zero PSA may result from some sort of optical aberration in the observer's eye, or because the observer's internal representation of space is non-veridical, or because the monitor display is distorted. Given these possibilities it makes more sense to measure the vernier threshold as the separation (or half the separation) between the points on the abscissa corresponding to the 0.25 and 0.75 response levels. This measure takes into account any bias. Alternatively, vernier acuity can be measured from the steepness, or slope, of the psychometric function. As we mentioned earlier, the slope of the psychometric function is inversely related to the standard deviation parameter of the function used to fit the data, so this standard deviation is a measure of vernier acuity (e.g., Watt & Morgan, 1983; McGraw et al., 2004). Recall also that the standard deviation parameter is a measure of precision, with a small standard deviation indicating a high precision. Whether the threshold or the slope is used as the basis of the measurement of vernier acuity, however, both are performance measures since the "better than" heuristic applies equally. Note, however, that because the PSA might be biased, it would be a mistake to treat the experiment as Type 1, i.e., one with a correct and an incorrect response on each trial. This is important. Suppose the observer's bias was such that when physically aligned, the upper line appeared slightly to the left of the lower line. On trials where the upper line was positioned slightly to the right of the lower line, the observer's bias would result in a number of "left" responses, and if feedback were given, such trials would be scored "incorrect." This would inevitably cause confusion in the observer – after all they really *did* see those lines as "left." Such confusion could be detrimental to performance.

The fact that a performance measure, vernier acuity, is in some circumstances best measured without feedback exemplifies how the performance versus appearance distinction is not always synonymous with Type 1 versus Type 2. In fact, precision, which we have argued is a performance measure, can be measured for any Type 2 PSE. Other examples of performance measures that are not necessarily obtained from Type 1 experiments are contrast detection thresholds obtained using the method of adjustment, measures of accuracy (see next paragraph) and measures of reaction time. Thus, although all Type 1 experiments measure performance, not all performance measures are obtained from Type 1 experiments.

Not only the precision but the bias in the vernier alignment task A can be considered to be a measure of performance. In Section 2.3.2 we defined the term accuracy as the closeness of a psychophysical measure to its true, i.e. physical value. For the vernier experiment, a bias indicates inaccurate alignment and thus the bigger the bias the lower the accuracy. A similar argument holds for the line bisection task illustrated in Figure 2.4. In this case, accuracy is measured by how close the observer's mean setting is to the physical mid-point of the line, while precision is related inversely to the variability of the observer's settings. Since one can legitimately argue that one observer is more accurate than another in vernier alignment or line bisection, accuracy as measured in these tasks is a performance measure. As a performance measure, accuracy is particularly germain to spatial vision where accurate estimates of distances and other spatial relationships are necessary in order for the observer to navigate the visual world. However, as we shall now see, measures of bias in many circumstances are better considered as measures of appearance.

Consider vernier alignment task B. In this case, as with the Muller–Lyer and brightness-matching experiments, it is the bias that we are primarily interested in. We want to know by how much the PSA is shifted by the presence of the white lines. The shift in the PSA is the separation between the PSAs measured in experiments A and B, with each PSA calculated as the point on the abscissa corresponding to 50% "left" responses. Assuming that the PSA from experiment A is at zero, the shift in PSA caused by the white lines is indicated by the green arrow on the graph associated with task B. This shift is a measure of appearance. All experiments that measure appearance are also Type 2.

Innumerable aspects of stimulus appearance are open to psychophysical measurement. To pick just four examples: choosing the computer sketch of a stimulus that best matches its appearance (e.g., Georgeson, 1992); indicating when a simulated three-dimensional random-dot rotating cylinder appears to reverse direction (e.g., Li & Kingdom, 1999); adjusting the colors of a moving chromatic grating until the grating appears to almost stop (Cavanagh, Tyler & Favreau, 1984); labeling contour-defined regions in images of natural scenes as being either "figure" or "ground" (e.g., Fowlkes, Martin, & Malik, 2007). Are there, however, any broad classes of procedure for measuring appearance? Matching and scaling are arguably two classes. Matching experiments measure PSEs between two physically different stimuli, as in the Rayleigh match, brightness-matching, Muller–Lyer, and vernier task B experiments described above. Scaling experiments, on the other hand, determine the relationship between the perceptual and physical representations of a stimulus, for example the relationship between perceived contrast and physical contrast, hue (or perceive chromaticity) and wavelength, perceived velocity and physical velocity, perceived depth and retinal disparity. Although not all perceptual scales are appearance-based, most are.

Example data from a scaling experiment are shown in Figure 2.9. Unlike the hypothetical data used so far to illustrate generic experimental results, every perceptual scale has a unique shape, so for Figure 2.9 we have reproduced a specific example from an experiment conducted by Whittle (1992). Whittle was interested in the relationship between the brightness (or perceived luminance) and physical luminance of uniform discs on a gray background. Observers were presented with a display consisting of 25 discs arranged in a spiral, with the first and last fixed in luminance at the lowest and highest available on the monitor – "black" and "white." The observer adjusted the luminances of the remaining 23 discs until they appeared to be at equal intervals in brightness. Figure 2.9 plots the number of the disc (1–25) against its luminance setting. If brightness (the perceptual dimension) was linearly related to luminance (the physical dimension) then the function would be a straight line. Instead, however, it has a complex shape. There are many different procedures

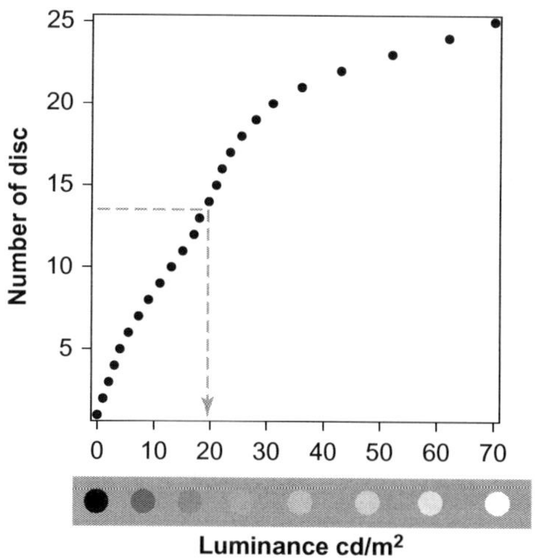

FIGURE 2.9 Data from a brightness scaling experiment. The graph plots the number of the disc against its luminance, after the luminances of all the discs have been adjusted to make them appear at equal brightness intervals. The green arrow indicates the point where the discs change from being decrements to increments. Data based on Whittle (1992).

for deriving perceptual scales, and these are summarized in Chapter 3, with further details provided in Chapter 7.

Both performance-based and appearance-based experiments are important to our understanding of vision. Measures from both types of experiment are probably necessary to fully characterize the system. The relationship between performance and appearance, and what each tells us about visual function, is a complex and interesting issue, but one beyond the remit of this book (e.g., in some instances they appear to measure closely-related psychological processes, as implied by Whittle, 1992, while in others they arguably measure quite different psychological processes, as argued by Gheorghiu & Kingdom, 2008) in relation to curvature processing. However, we argue that the performance-versus-appearance dichotomy more than any other so far discussed represents the most fundamental dividing line in psychophysics. For this reason we propose it as the candidate for the superordinate division in our classification scheme. In the next section, we discuss some of the possibilities for the second-level categories in the scheme.

2.3.5 "Forced-choice" Versus "Non-forced-choice"

Forced-choice procedures are used extensively in psychophysics. There is more than one convention, however, for the use of the term "forced-choice." In signal detection theory (McNicol, 2004; Macmillan & Creelman, 2005; Wickens, 2002), discussed in Chapter 6, the term is mainly used to characterize experiments in which two or more stimulus alternatives are presented during a trial, one of which is the "target." Example forced-choice tasks that accord with this usage are: deciding which of two stimuli, a blank field or patch, contains the patch; deciding which of two patches is brighter; deciding which of three lines, two oriented −5 degrees and one oriented +5 degrees, is the −5 degree line. In each of these examples, the observer has to select a stimulus from two or more presented during the trial. Typically, at the end of the experiment the proportion of trials in which the target alternative was selected over the other(s) is calculated for each stimulus magnitude. Recall that the measure derived from these proportions may be a performance measure, such as a threshold, or an appearance measure such as a PSE.

In the signal detection literature, most other types of discrimination task are not explicitly referred to as forced-choice, we understand to avoid the term becoming redundant. For example, take the procedure termed "yes/no," in which only one stimulus is presented per trial. Figure 2.10 illustrates the procedure when applied to a contrast detection threshold experiment, along with the two-stimulus-per-trial version (2AFC) explicitly referred to as forced-choice in the signal detection literature. In the yes/no experiment, the target is typically presented on half the trials and the observer responds "yes" or "no" on each trial depending on whether or not they see the target. Although yes/no experiments figure prominently in the signal detection

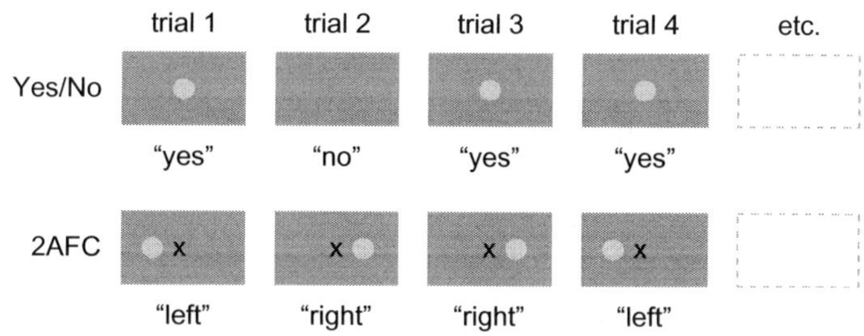

FIGURE 2.10 Yes/no versus 2AFC (two-alternative forced-choice) procedures. In the yes/no task the two alternatives – "stimulus present" and "stimulus absent" – are presented on separate trials, whereas in the 2AFC task they are presented within the same trial. Correct responses are indicated below the stimuli. In this book, both types of task are referred to as "forced-choice."

literature, they are not widely employed today in visual psychophysics, as the 2AFC procedure is generally preferred for reasons discussed later and in Chapters 3 and 6. The more popular type of single-stimulus-per-trial experiment is the variety we term "symmetric," meaning that the stimulus alternatives are akin, metaphorically, to mirror images, i.e., they are "equal and opposite." Example symmetric one-stimulus-per-trial experiments include the orientation discrimination task illustrated in Figure 2.4 (grating left oblique versus grating right oblique) and the vernier task A in Figure 2.8 (upper line to the left versus upper line to the right). Although in the vernier alignment experiment two lines are presented to the observer on each trial, one should still think of this as a single stimulus alternative. As with the yes/no task, signal detection theory does not generally refer to symmetric single-stimulus-per-trial experiments as forced-choice.

We argue however that in the context of psychophysics as a whole it is important to distinguish between procedures that require forced-choice responses and procedures that do not. Therefore, in this book, we have adopted the convention of referring to any procedure as forced-choice if the observer has two or more pre-specified response options. According to this convention, a yes/no experiment is forced-choice because there are two response options: "yes" and "no." By the same argument, the single-alternative-per-trial orientation-discrimination and vernier acuity experiments described above are also forced-choice. We refer to this convention as the response-based definition of forced-choice. Readers may prefer to think of the response-based definition of forced-choice in terms of choices between "stimulus states," for example in the yes/no experiment between "stimulus present" and "stimulus absent". As it turns out, the response-based definition of forced-choice is widely adopted in both the literature and in common parlance, as exemplified by the many single-stimulus-per-trial experiments that are routinely termed forced-choice (e.g., Dakin, Williams, & Hess, 1999).

Are there drawbacks to a response-based definition of forced-choice? Consider the method of limits, used mainly to obtain thresholds. Observers are presented with a series of stimuli that are systematically increased (or decreased) in intensity, and are prompted to indicate whether or not they can see the stimulus. The stimulus intensity at which the observer switches response from "no" to "yes" (or *vice versa*) is then taken as the threshold. With a response-based definition of forced-choice, the procedure is arguably forced-choice. Suppose, however, the observer "takes control" of the stimulus presentation and adjusts the stimulus himself/herself. This is normally regarded as the method of adjustment and not forced-choice. But are the two procedures really so different? In both experimenter-controlled and observer-controlled procedures there is no correct and incorrect answer on each stimulus presentation, because the stimulus is always present, albeit with different intensity, so both procedures are Type 2. Moreover, with the observer-controlled adjustment procedure the observer is constantly making a perceptual decision as to whether or not the stimulus is visible, so is this not forced-choice according to our definition? The example of the method of limits highlights a conundrum for the response-based definition of forced-choice: where does forced-choice end and method of adjustment begin? The resolution of the conundrum lies in a caveat to our definition of forced-choice, namely that the experiment must involve clearly demarcated trials.

Forced-choice tasks are invariably denoted by the abbreviations AFC (alternative forced-choice) or IFC (interval forced-choice). AFC is the generic term, while IFC is reserved for procedures in which the stimulus alternatives are presented in temporal order. Both acronyms are invariably prefixed by a number. In this book, this number is the number of stimulus alternatives presented on each trial, denoted by M. The value of M is important for the signal detection analyses described in Chapter 6, since it relates to the degree of uncertainty as to the target interval/location, as well as to the amount of information present during a trial. Because we have adopted the convention of characterizing all tasks that require forced-choice responses as AFC or IFC, we characterize single-stimulus-per-trial procedures such as the yes/no and symmetric single-interval tasks as 1AFC. To spell out our usage: 1AFC means "… a forced-choice task in which only one stimulus alternative is presented per trial." Readers should be aware, however, that other investigators use the number of response choices as the prefix when referring to single-stimulus-per-trial experiments, which is typically 2 (e.g., Dakin et al., 1999).

We denote on the other hand the number of response choices by m. In most procedures M and m are the same. For example, in tasks where one interval contains the target and the other a blank field there are two alternatives per trial – blank field and target – and two response choices per trial – "1" (first interval) and "2" (second interval). However, with a single-interval task with two response choices, the use of m as the prefix leads to the notation 2AFC (e.g., as mentioned above for Dakin et al., 1999), rather than the 1AFC notation as employed here. In most other cases the use of the prefix M accords with current convention, e.g., two-alternative tasks are also here 2AFC, two-interval tasks are 2IFC, three-alternative tasks are 3AFC, and so on.

Our choice of M rather than m as the prefix for a forced-choice task is a concession to signal detection theory, where the distinction between single-interval/alternative and two-interval/alternative tasks needs to be explicit. Nevertheless, m is an important parameter as it determines the guessing rate in a forced-choice task. The guessing rate is the proportion of times an observer is expected to be correct if simply guessing, and is hence calculated as $1/m$, for example 0.5 in both a yes/no and 2AFC task. The guessing rate is a critical parameter when fitting psychometric functions, as we shall see in Chapter 4.

A third important parameter in forced-choice tasks is the number of stimuli presented per trial, denoted here by N. Again, in most procedures N is the same as M (and hence m). However, in some forced-choice tasks, such as the "same-different" task that will be discussed in more detail in Chapters 3 and 6, the values of N and M are not the same. Same-different tasks in vision research typically use either two or four stimuli per trial, i.e., N is 2 or 4. In the $N = 2$ version, the two stimuli on each trial are either the same or are different, and the observer is required to respond "same" or "different." In the $N = 4$ version, a same pair *and* a different pair are presented in each trial, usually in temporal order, and the observer responds "1" or "2" depending on the interval perceived to contain the same (or different) pair. In both the $N = 2$ and $N = 4$ same-different versions, m, the number of response alternatives, is 2. M, the number of stimulus alternatives per trial, is, respectively 1 and 2. Values of N, m and M for a variety of different psychophysical tasks are given in Table 6.1 in Chapter 6.

2.3.6 "Criterion-free" Versus "Criterion-dependent"

The yes/no task described above is often termed "criterion-dependent," whereas the 2AFC/2IFC task is often termed "criterion-free." Characterizing yes/no tasks as criterion-dependent captures the fact that observers typically adopt different criteria as to how strong the internal signal must be before they respond "yes." Different criteria lead to different biases towards "yes" or "no," irrespective of the actual strength of the internal signal. If a strict criterion is adopted, the internal signal must be relatively strong for the observer to respond "yes," whereas if a loose criterion is adopted a weak internal signal is sufficient. The adoption of different criteria might result from an unconscious bias, or it might be part of a conscious strategy. For example, observers might consciously bias their responses towards "yes" because they want to maximize the number of correct target detections or "hits," even if this results in a number of "false alarms," i.e., "yes" responses when the target is absent. On the other hand, they might consciously adopt a strict criterion in order to minimize the number of false alarms, even if this means fewer hits.

2AFC/2IFC tasks can also be prone to bias, but a bias towards responding "1" (first alternative/interval) or towards "2" (second alternative/interval). However, biases of this sort are less common because the two response choices are on an "equal footing;" the observer knows that on every trial the target will be present, so the option of consciously trading off hits and false alarms does not in the same way arise. When biases do occur in forced-choice tasks one cannot estimate the sensitivity of an observer to the target stimulus from the proportion correct responses. Chapter 6 explains why this is so, and describes an alternative measure, d' ("d-prime"), that is more valid in such cases.

There is, however, another more general meaning to the terms criterion-free and criterion-dependent. Occasionally one hears that Type 1 tasks are criterion-free and Type 2 tasks criterion-dependent. This usage somewhat parallels the objective–subjective dichotomy discussed above.

2.3.7 "Detection" Versus "Discrimination"

The terms "detection" and "discrimination" are used variously to characterize tasks, measures, procedures, and experiments. For example one might carry out a "detection experiment" using a "detection task" to obtain a "detection measure." The term detection is most frequently used to characterize experiments that measure thresholds for detecting the presence, as opposed to the absence, of a stimulus, for example a contrast "detection" threshold. However, the "null" stimulus in a detection experiment is not necessarily a blank field. In curvature detection experiments the null stimulus is a straight line, as illustrated at the top of Figure 2.11. Similarly, in stereoscopic depth detection experiments the null stimulus lies at a depth of zero, i.e., in the fixation plane, and in a motion detection experiment the null stimulus is one that is stationary.

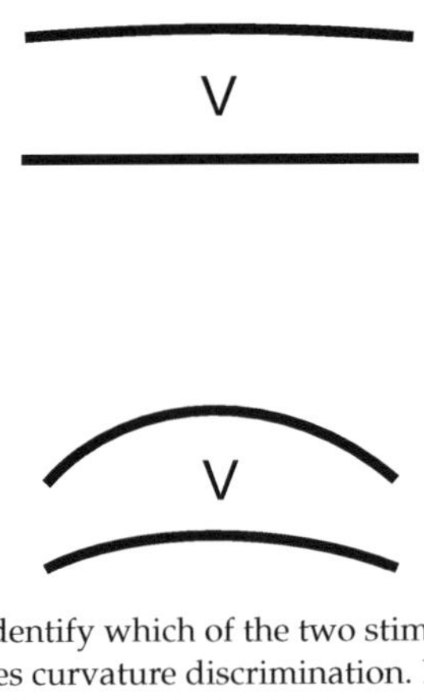

FIGURE 2.11 Top: the task is to identify which of the two stimuli is curved. The task is sometimes termed curvature detection, sometimes curvature discrimination. Bottom: the task is to identify which stimulus is the more curved. This task is invariably termed curvature discrimination.

The term discrimination, on the other hand, is generally reserved for experiments in which neither of the two discriminands (the stimuli being discriminated) is a null stimulus. Thus, in a curvature discrimination experiment, illustrated at the bottom of Figure 2.11, both stimuli in the forced-choice pair are curved, and the task is to decide which is more curved. Similarly, in a stereoscopic depth discrimination experiment both stimuli have non-zero depth, and the task is to decide which is nearer (or further) and in a motion discrimination experiment both stimuli are moving, and the task is to decide which is moving faster (or slower).

This being said, the terms detection and discrimination tend to be interchangeable. For example, the curvature task illustrated at the top of Figure 2.11 is sometimes termed detection (Kramer & Fahle, 1996) and sometimes discrimination (e.g., Watt & Andrews, 1982), even though one of the discriminands is a straight line. Consider also the contrast discrimination experiment illustrated in Figure 2.12. The aim here is to measure the just-noticeable difference (JND) in contrast between two above-threshold stimuli. Typically, one of the contrasts, say the one on the left in the figure, is fixed and termed the baseline or pedestal contrast. The other stimulus is varied to find the JND. One can think of this experiment in two ways. On the one hand it measures a "discrimination" threshold between two contrasts, while on the other hand it measures a "detection" threshold for an increment in contrast added to a pedestal. In Figure 2.12 the pedestal and pedestal-plus-increment are presented to the observer at the same time, termed by some the "pulsed-pedestal" paradigm (e.g., Lutze, Pokorny, & Smith, 2006). In another form of the procedure the pedestals are first presented together, and then after a short duration the increment is added to one of the pedestals, termed by some the "steady-pedestal" paradigm

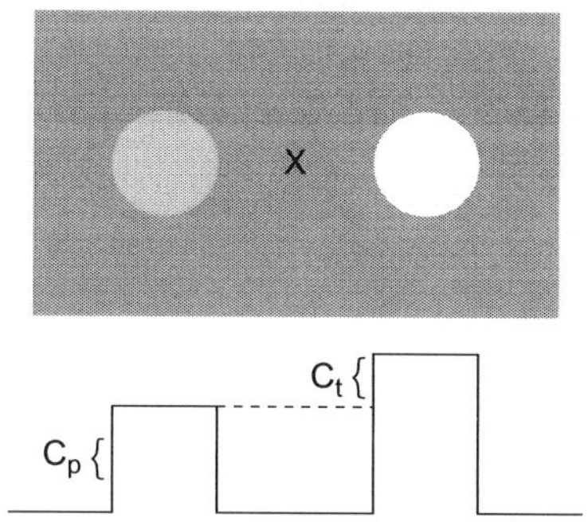

FIGURE 2.12 The task of the subject is to indicate the patch with the higher contrast. The lower contrast patch on the left is fixed in contrast and is termed the pedestal contrast Cp. The variable contrast patch is the one on the right. The task can be regarded as either contrast "discrimination" or contrast increment "detection." The contrast increment is the test contrast, Ct.

(e.g., Lutze et al., 2006). One could make the argument that the two paradigms should be considered respectively discrimination and detection, but in reality there is no hard-and-fast rule here and both paradigms could be considered detection or discrimination. One should be prepared to be flexible in the use of these terms.

Two psychophysical terms closely related to detection and discrimination are "recognition" and "identification." "Recognition" generally denotes experiments involving relatively complex stimuli such as faces, animals, and household objects, where the task is to select from two or more objects one either recently shown or long ago memorized. For example, in a prototypical face-recognition experiment a briefly-presented test face is followed by two or more comparison faces from which the observer must choose the test face (e.g., Wilbraham et al., 2008). This type of procedure is known as "match-to-sample." Another type of face recognition task requires the observer to simply name a briefly-presented famous face (e.g., Reddy, Reddy, & Koch, 2006).

The term "identification" is sometimes used instead of recognition, sometimes instead of discrimination. Probably the most common usage of the term is to characterize experiments in which the discriminands differ along two dimensions, both of which must be discriminated. For example, in a type of experiment termed "simultaneous detection and identification" the observer is presented with two intervals on each trial (i.e., 2IFC), one containing the target and the other a blank field. However, the target can be one of two types of stimulus, e.g., red or green, moving left or moving right, near or far. The observer is required to make two judgements on each trial: one the interval containing the stimulus and the other the type of stimulus. The first judgement is usually termed detection, while the second is sometimes termed discrimination (e.g., Watson & Robson, 1981) and sometimes identification (e.g., Kingdom & Simmons, 1998). Typically, the aim of the experiment is to decide whether the psychometric functions derived from the two types of decision are significantly different (see Chapter 8 for details).

2.3.8 "Threshold" Versus "Suprathreshold"

As with the terms detection and discrimination, "threshold" and "suprathreshold" can refer to experiments, tasks, procedures, or measures. In sensory science a threshold is roughly defined as the stimulus magnitude that results in the perception of a new stimulus state. Traditionally, psychophysical thresholds have been divided into two categories: "absolute" and "difference." An absolute threshold is the magnitude of a stimulus that enables it to be just discriminated from its null, as exemplified by a contrast detection threshold (Figure 2.12). A difference threshold, on the other hand, is the magnitude of stimulus difference needed to make two stimuli that are both above their individual absolute thresholds just discriminable, as exemplified by a contrast discrimination threshold (Figure 2.12).

Both these examples of threshold measures are performance measures. However, not all thresholds are performance measures. Consider the phenomenon of binocular rivalry. Binocular rivalry is said to occur when different stimuli presented to the two eyes are perceived to alternate in dominance (e.g., Papathomas, Kovacs, & Conway, 2005). A threshold for binocular rivalry can be defined as the minimum physical difference between the stimuli needed to produce rivalry. This is an appearance measure.

The term suprathreshold has more than one definition. One definition is that it is any non-threshold experiment, task, procedure, or measure. According to this definition the contrast-matching and Muller–Lyer experiments described above are suprathreshold, but the contrast discrimination, vernier acuity, and curvature discrimination experiments are not, because they measure thresholds. However, suprathreshold can also refer to any experiment/task/procedure/measure that involves stimuli that are all individually above their own detection threshold. According to this definition the contrast discrimination, vernier acuity and curvature discrimination experiments are also suprathreshold. Once again, one has to be prepared to be flexible when interpreting these terms.

2.4 CLASSIFICATION SCHEME

The first four levels of our proposed scheme are illustrated in Figure 2.13; a fifth level is added in the next chapter. Let us recap the meaning of these categories. Any experiment,

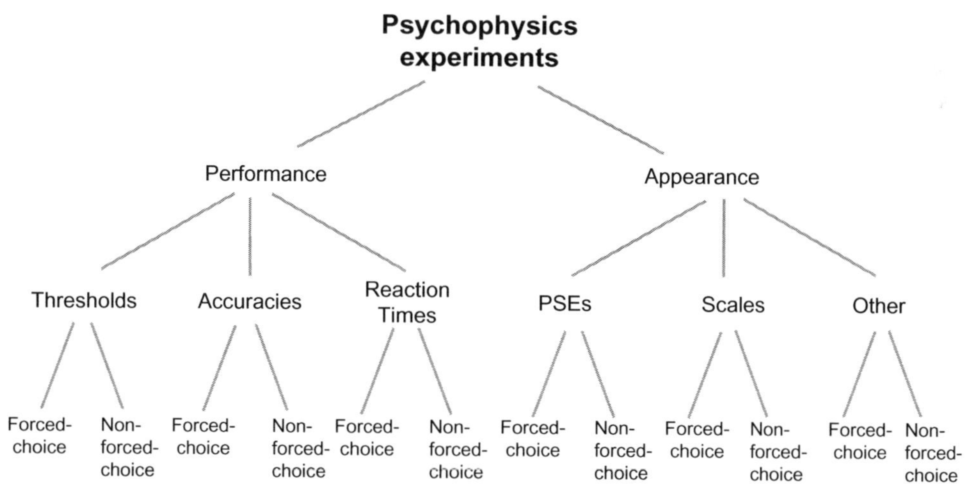

FIGURE 2.13 The initial stages of a scheme based on the performance-appearance distinction. An expanded version of the scheme is provided in the following chapter.

task or procedure is performance-based if what it measures affords a comparison in terms of aptitude. Thus, a contrast-detection experiment is performance-based, because it affords the claim that contrast sensitivity is better in central compared to peripheral vision. Similarly, vernier acuity is a performance measure because it affords the claim that vernier acuity is better in the young than in the old, and speed discrimination is a performance measure because it affords the claim that speed discrimination is better at low than at high speeds. Appearance-based experiments, on the other hand, measure the apparent magnitude of some stimulus dimensions. Thus, an experiment that measures the Muller–Lyer illusion measures the apparent difference in line length between the two figures, while the asymmetric brightness-matching experiment measures the apparent brightness of a patch surrounded by an annulus. Given that the same task can be used to obtain both performance and appearance measures, the performance-versus-appearance dichotomy speaks primarily to the "goal" of a psychophysical experiment and the "measure" it provides. We regard performance and appearance measures of sensory function as equally important to our understanding of sensory processes, and in the rest of the book we have attempted to balance their respective treatments.

Thresholds (which here include precisions) are the best-known performance measures, but performance measures also include proportion correct, d's (d-primes), measures of accuracy and reaction times. The most common appearance-based measures are PSEs (derived from matching procedures) and perceptual scales (derived from scaling procedures). Therefore, the third level in the scheme highlights thresholds, accuracies, reaction times, PSEs and scales.

The fourth-level division into forced-choice and non-forced-choice is intended to shift the emphasis of the scheme from the measurement goal of a psychophysical experiment to its procedural form. In the next chapter a fifth level is added, a division by the number of stimuli presented per trial, providing the final framework for systematically examining a wide range of psychophysical procedures.

Further Reading

A discussion of Brindley's distinction between Class A and Class B observations can be found in Brindley (1970) and Gescheider (1997). Sperling has written a short guide to Type 1 and Type 2 (Sperling, 2008), although see also Galvin et al. (2003) for a somewhat different interpretation. Discussions of yes/no versus forced-choice procedures from the standpoint of signal-detection theory can be found in McNicol (2004), MacMillan & Creelman (2005), and Wickens (2002). A good example of the congruency of threshold and scaling measures can be found in Whittle (1992), while a discussion of the incongruency between performance and appearance measures can be found in the discussion of studies of curvature perception in the introduction of Gheorghiu & Kingdom (2008).

Exercises

1. Categorize the following observations as Class A or Class B.
   a. Choosing a previously shown face from a set of five alternatives (a match-to-sample face recognition task).
   b. Deciding whether a particular purple is more reddish or more bluish.
   c. Measuring the effect of contrast on the perceived speed of a moving object.
   d. Measuring the just-noticeable-difference between the lengths of two lines.
   e. Naming a briefly-presented famous face.
   f. Measuring the reaction time to the onset of a grating.
   g. Measuring the threshold for identifying that an image of an everyday scene has been artificially stretched.
   h. Measuring the duration of the motion-after-effect (the illusory reversed motion seen in an object following adaptation to a moving object).
2. Which of the following could be measured using a Type 1 forced-choice task (i.e. with a correct and an incorrect response on each trial)?
   a. Estimating the perceived speed of a moving pattern.
   b. Bisecting a line into two equal halves.
   c. Deciding whether a particular purple is more reddish or more bluish.
   d. Measuring the just-noticeable-difference between the curvature of two lines.
   e. Discriminating male from female faces.
3. Make a table with nine rows labeled by the dichotomies described in the chapter and six columns a–f. For each of the following tasks, consider which term from each dichotomy, if at all, is appropriate and write the answer in the table.
   a. The observer adjusts the contrast of a patch until it looks just-noticeably-brighter than another patch.
   b. The observer presses a button in response to a decremental change in contrast and his/her reaction time is measured.
   c. The observer chooses from two colors the one appearing more yellowish.
   d. The observer adjusts the speed of a drifting grating until it matches the perceived speed of another drifting grating with a different spatial frequency (the spatial frequency of a grating is the number of cycles of the grating per unit visual angle).
   e. The observer selects on each trial which of two depth targets appears to lie in front of the fixation plane.
   f. The observer identifies whether the face presented on each trial is male or female.

References

Papathomas, T. V., Kovacs, I., & Conway, T. (2005). Interocular grouping in binocular rivalry: Basic Attributes and combinations Chapter 9. In D. Alais & R. Blake (Eds.), *Binocular Rivalry*. Cambridge, MA: MIT Press.

Blakeslee, B., & McCourt, M. E. (1997). Similar mechanisms underlie simultaneous brightness contrast and grating induction. *Vision Research*, *37*, 2849–2869.

Brindley, G. S. (1970). *Physiology of the Retina and Visual Pathway*. Baltimore: Williams and Wilkens.

Cavanagh, P., Tyler, C. W., & Favreau, O. E. (1984). Perceived velocity of moving chromatic gratings. *Journal of the Optical Society of America A, 1*, 893–899.

Dakin, S. C. (2001). Information limit on the spatial integration of local orientation signals. *Journal of the Optical Society of America A, 18*, 1016–1026.

Dakin, S. C., Williams, C. B., & Hess, R. F. (1999). The interaction of first- and second-order cues to orientation. *Vision Research, 39*, 2867–2884.

Fowlkes, C. C., Martin, D. R., & Malik, J. (2007). Local figure-ground cues are valid for natural images. *Journal of Vision, 7*(8), 1–9.

Macmillan, N. A., & Creelman, C. D. (2005). *Detection Theory: A User's Guide*. Mahwah, NJ: Lawrence Erlbaum Associates.

McNicol, D. (2004). *A Primer of Signal Detection Theory*. Mahwah, NJ: Lawrence Erlbaum Associates.

Morgan, M. J., Hole, G. J., & Glennerster, A. (1990). Biases and sensitivities in geometric illusions. *Vision Research, 30*, 1793–1810.

Gescheider, G. A. (1997). *Psychophysics: The Fundamentals*. Mahwah, NJ: Lawrence Erlbaum Associates.

Galvin, S. J., Podd, J. V., Drga, V., & Whitmore, V. (2003). Type 2 tasks in the theory of signal detectability: Discrimination between correct and incorrect decisions. *Psychonomic Bulletin and Review, 10*, 843–876.

Georgeson, M. A. (1992). Human vision combines oriented filters to compute edges. *Proceedings of the Royal Society B, 249*, 235–245.

Gheorghiu, E., & Kingdom, F. A. A. (2008). Spatial properties of curvature-encoding mechanisms revealed through the shape-frequency and shape-amplitude after-effects. *Vision Research, 48*, 1107–1124.

Hong, S. W., & Shevell, S. K. (2004). Brightness induction: Unequal spatial integration with increments and decrements. *Visual Neuroscience, 21*, 353–357.

Kramer, D., & Fahle, M. (1996). A simple mechanism for detecting low curvatures. *Vision Research, 36*(10), 1411–1419.

Kingdom, F. A. A., & Simmons, D. R. (1998). The missing-fundamental illusion at isoluminance. *Perception, 27*, 1451–1460.

Li, H-C. O., & Kingdom, F. A. A. (1999). Feature specific segmentation in perceived structure-from-motion. *Vision Research, 39*, 881–886.

Lutze, M., Pokorny, J., & Smith, V. C. (2006). Achromatic parvocellular contrast gain in normal and color defective observers: Implications for the evolution of color vision. *Visual Neuroscience, 23*, 611–616.

McGraw, P. V., McKeefry, D. J., Whitaker, D., & Vakrou, C. (2004). Positional adaptation reveals multiple chromatic mechanisms in human vision. *Journal of Vision, 4*, 626–636.

Rayleigh, L. (1881). Experiments on colour. *Nature, 25*, 64–66.

Reddy, L., Reddy, L., & Koch, C. (2006). Face identification in the near-absence of focal attention. *Vision Research, 46*, 2336–2343.

Shevell, S. K., Sun, Y., & Neitz, M. (2008). Protanomaly without darkened red is deuteranopia with rods. *Vision Research, 48*, 2599–2603.

Sperling, G.B. (2008). Type I and Type II experiments. http://aris.ss.uci.edu/HIPLab/ProSem202c/UCI_access/READINGS/Type_1_and_Type_2_Expts.pdf

Sperling, G., Dosher, B. A., & Landy, M. S. (1990). How to study the kinetic depth experimentally. *Journal of Experimental Psychology: Human Perception and Performance, 16*, 445–450.

Thomas, P. B., & Mollon, J. D. (2004). Modelling the Rayleigh match. *Visual Neuroscience, 21*, 477–482.

Watson, A. B., & Robson, J. C. (1981). Discrimination at threshold: labeled detectors in human vision. *Vision Research, 21*, 1115–1122.

Watt, R. J., & Andrews, D. P. (1982). Contour curvature analysis: hyperacuities in the discrimination of detailed shape. *Vision Research, 22*, 449–460.

Watt, R. J., & Morgan, M. J. (1983). The use of different cues in vernier acuity. *Vision Research*, *23*, 991–995.
Wickens, T. D. (2002). *Elementary Signal Detection Theory.* Oxford, New York: Oxford University Press.
Whittle, P. (1992). Brightness, discriminability and the "crispening effect.". *Vision Research*, *32*, 1493–1507.
Wilbraham, D. A., Christensen, J. C., Martinez, A. M., & Todd, J. T. (2008). Can low level image differences account for the ability of human observers to discriminate facial identity? 5, 1–12. *Journal of Vision*, *8*(15).

CHAPTER

3

Varieties of Psychophysical Procedure

3.1 INTRODUCTION

In this chapter we delve into the design details of psychophysical procedures and consider their relative advantages and disadvantages. The term "procedure" refers to both the observer's task and the method of data collection, in other words the "front-end" of a psychophysics experiment. Subsequent chapters will deal with the analysis of psychophysical data – the "back-end" of a psychophysics experiment. The procedures described in this chapter are organized according to the performance versus appearance classification scheme described in Chapter 2. Figure 3.1 expands the scheme to include a further level of categorization based on the number of stimuli presented per trial, or *N*. *N* seems to us the natural way to extend the scheme in order to incorporate the many variants of each class of psychophysical procedure.

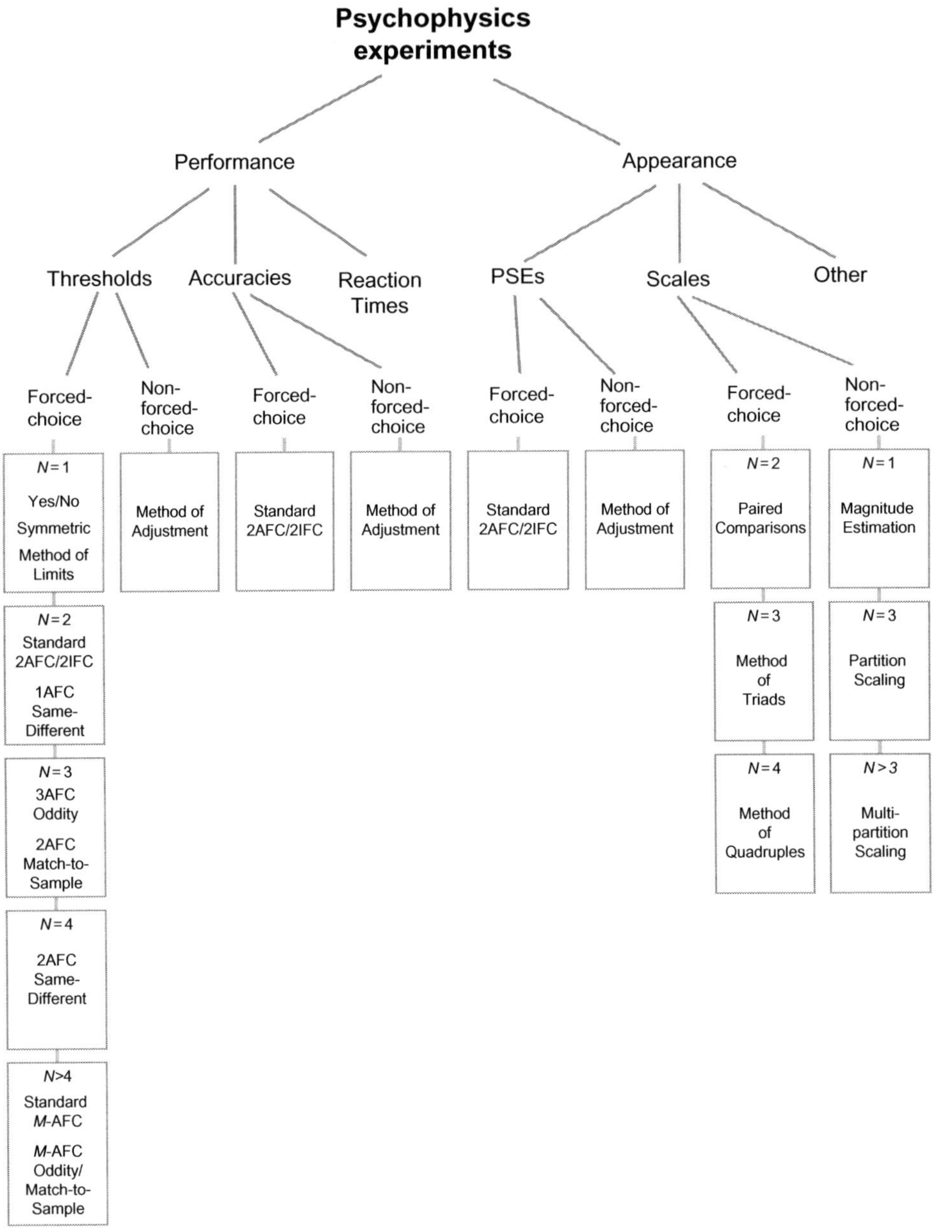

FIGURE 3.1 Expanded scheme for classifying psychophysical experiments.

Recall the meaning of the major categories in the classification scheme. Performance-based procedures measure aptitude or "how good one is" at a particular visual task. In Chapter 2 we included in this category most types of threshold measure, simple proportion correct, d' measures of precision, accuracies and reaction times. We pointed out that while many types of performance measure could be obtained using Type 1 tasks, i.e., tasks with a correct and an incorrect response on each trial, not all were Type 1. For example, measures of reaction time, some measures of precision, accuracies and some types of threshold measure could be obtained using procedures that were Type 2, i.e., had no correct and incorrect response.

Appearance-based procedures, on the other hand, generally measure the apparent magnitude (relative or absolute) of some stimulus dimension. Appearance-based procedures can only be Type 2. This does not imply that appearance-based measurements are less useful or less valid than performance-based measurements for understanding sensory function. Both types of measurement are arguably necessary to fully characterize the system.

Forced-choice procedures, as defined here, refer to procedures in which the observer is required on each trial to make a response from two or more pre-specified options. Our definition of forced-choice is not restricted to situations where two or more stimuli are presented per trial; a single-stimulus-per-trial presentation with two response options is here regarded as forced-choice. Moreover, the term forced-choice applies to both performance-based and appearance-based procedures.

3.2 PERFORMANCE-BASED PROCEDURES

3.2.1 Thresholds

3.2.1.1 Forced-choice Threshold Procedures

Although we have chosen to categorize forced-choice procedures according to the number of stimuli presented per trial, N, recall from Chapter 2 that the acronyms AFC and IFC are not prefixed by N, but M, the number of stimulus alternatives presented per trial. In many types of forced-choice procedure, N and M are the same, but in some, such as the same-different and match-to-sample procedures discussed later, they are different.

Consider the various ways one might measure an orientation discrimination threshold for grating patches. The goal is to measure the minimum discriminable difference in orientation between a left-oblique and a right-oblique patch of grating. The first thing to note is that the potential number of stimuli that could be presented during a trial is infinite. For example, the display screen could be divided into 100 squares by an 11×11 grid of lines, with 99 locations containing, say, a left-oblique grating and 1 location containing the right-oblique "target" grating. The task for the observer would be to choose the location containing the target, and the response would be scored as either correct or incorrect. In principle this seems fine, but think about what the task would involve. During each trial the observer would need to scan all 100 locations in order to be sure not to miss the target. Therefore, each trial would invariable take several seconds or longer. The experiment would take a long time to complete, assuming one wants to collect enough data to obtain a good estimate of the threshold, say from a psychometric function of proportion correct versus orientation difference. The procedure seems impractical, unless of course one specifically wants to study how observers perform with large N displays. For most purposes, however, a small value of N is preferable. In the limit $N = 1$, but there can be drawbacks to $N = 1$. In fact, as we shall see, there are both advantages and disadvantages to each of $N = 1, 2, 3, 4$, and $N > 4$.

Figure 3.2 summarizes the common varieties of performance-based forced-choice tasks using small N, applied to the orientation discrimination experiment. We will discuss the various options illustrated in the figure as we proceed through this section. Note how the value of M, which prefixes the acronym AFC, is not always the same as N.

3.2.1.1.1 $N = 1$ (one stimulus per trial)

Method of limits

Although rarely used these days, this procedure is a quick method of obtaining a rough estimate of a threshold. It is probably most useful for getting a handle on the appropriate stimulus levels to use in subsequent more rigorous experiments. The method of limits may also be desirable in situations where the experimenter needs to maintain close verbal contact with the observer. A verbal report may be the only possible type of response with for example, young children or clinically impaired persons, or indeed in any circumstance where it is difficult for the observer to be "in the driving seat."

The observer is presented with a series of temporally or spatially demarcated stimuli of increasing (ascending method of limits) or decreasing (descending method of limits) magnitude, including sometimes the null or baseline stimulus at one end of the continuum. In a contrast detection threshold experiment, the ascending series might be contrasts of, say, 0, 0.01, 0.02, 0.04, 0.08, etc. For our orientation discrimination example the series might be grating patch orientations of, say, 0, 0.25, 0.5, 0.75, 1.0, 1.25, etc., degrees. On each presentation the observer is required to report "yes" or "no," depending on whether the stimulus appears noticeably different from the null or baseline level (zero in both examples). The threshold in each case is measured as the stimulus magnitude at which the response switches from "no" to "yes" and/or *vice versa*. This is a Type 2 performance procedure, because the observer's response is never evaluated in terms of whether it is correct or incorrect. Typically, the ascending and descending series are presented alternately, and the thresholds from each averaged.

A potential disadvantage of the method of limits is that the observer may become accustomed to reporting that they perceive (or not) a stimulus, and as a result

N	Task name	Stimuli presented during trial	Task
1	1AFC Symmetric	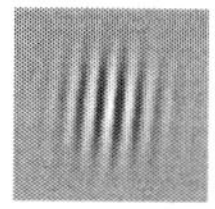	Respond "left-oblique" or "right-oblique"
2	Standard 2AFC	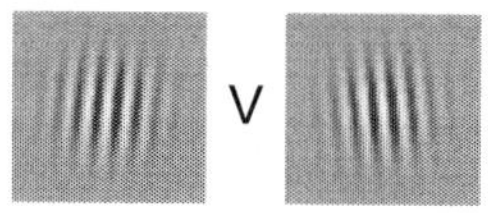	Select stimulus that is left-oblique
2	1AFC Same-Different	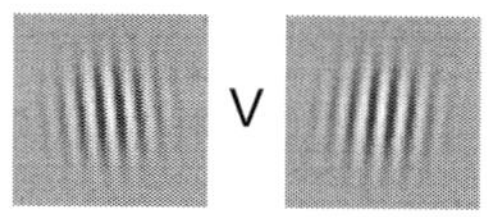	Respond "same" or "different"
3	3AFC Oddity	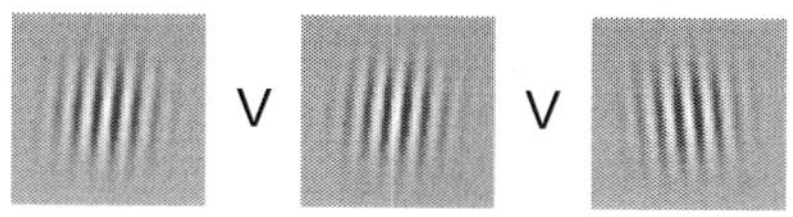	Select stimulus that is the oddity
3	2AFC Match-to-sample	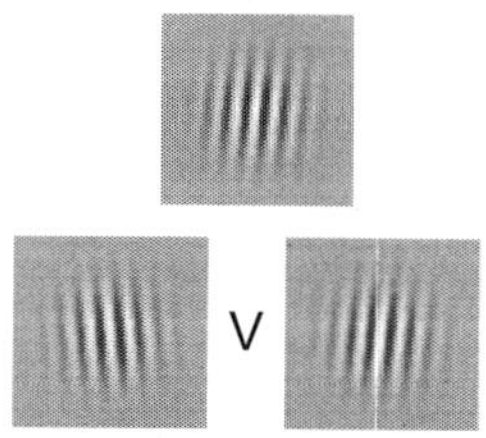	Select from the two bottom stimuli the one that is the same as the top stimulus
4	2AFC Same-Different	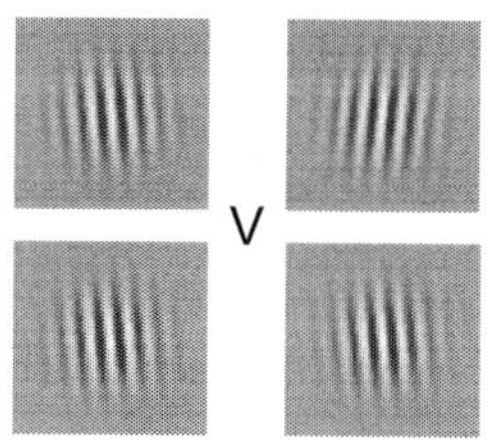	Select the pair (top or bottom) that is different (or same)

FIGURE 3.2 Different methods for measuring an orientation discrimination threshold. N = number of stimuli presented on each trial. Note that the number that prefixes the acronym AFC (alternative-forced-choice) is M, the number of stimulus alternatives presented per trial.

continue to give the same report even at stimulus magnitudes that are higher (or lower) than the "real" threshold. This is termed the error of habituation. Conversely, the observer may anticipate that the stimulus is about to become detectable, or undetectable, and make a premature judgement. This is called the error of expectation.

Errors due to habituation and expectation may be minimized by averaging thresholds from ascending and descending series.

Yes/No

The yes/no procedure is employed primarily for measuring detection thresholds. Typically, half the trials contain the target stimulus and half no target stimulus, and the task for the observer is to respond "yes" or "no" on each trial. Since the responses are evaluated as either correct or incorrect, the procedure is Type 1. As in all forced-choice tasks, the order of presentation of the target-present and target-absent trials must be random, or quasi-random. With quasi-random presentation a rule precludes long sequences of either target-present or target-absent trials.

Yes/no tasks are particularly prone to the effects of bias. The bias in this case means that observers may adopt, intentionally or unintentionally, different criteria as to how much sensory evidence they require before being prepared to give a "yes" response. If they adopt a strict criterion, they will respond "yes" only on those trials when they are very confident that the target is present. On the other hand if they adopt a loose criterion, they will respond "yes" on the flimsiest of evidence. Experimenters sometimes use the yes/no task *because* it is criterion-dependent, for example in order to study the effect of incentives on performance. The incentive might be to maximize the number of "hits" – these are "yes" responses when the target is present, or minimize the number of "false alarms" – these are "yes" responses when the target is absent. In situations where the observer is biased towards either responding "yes" or "no," the proportion of correct decisions is a poor measure of how sensitive the observer is to the target stimulus. To circumvent this problem, experimenters typically prefer the measure of performance d' that can be calculated from the proportions of hits and false alarms. The method for doing this is discussed in Chapter 6.

Symmetric

$N = 1$ forced-choice procedures can also be used when the two discriminands are "symmetric," as in the orientation discrimination task illustrated at the top of Figure 3.2. Here, the two discriminands are left- and right-oblique grating patches, but of course only one is presented during each trial. Because the two discriminands are "equal but opposite," it is less likely that observers will be biased towards responding to one more than the other. Hence, many experimenters treat symmetric $N = 1$ tasks as "bias-free" and use proportion correct as the measure of performance. However, to be sure, one should analyze the data in a similar way to the yes/no task, as described in Chapter 6. To minimize the possibility of bias it is important to make observers aware that the discriminands are presented an equal number of times, or with equal probability.

The main advantages of the symmetric $N = 1$ task is that a large number of responses can be collected within a relatively short time, and the task has minimum cognitive load. Inexperienced observers often find this task one of the easiest. Typically, the experimenter presents the observer with different stimulus magnitudes during an experimental session, either by the method of constants or by an adaptive procedure (see below).

3.2.1.1.2 $N = 2$

Standard 2AFC/2IFC

In what is probably the most popular design in performance-based forced-choice psychophysics, observers are presented on each trial with two stimuli, and are required to select one as the target (Figure 3.2). In one form of the task the two stimuli are presented together on the screen (2AFC), while in the other they are presented in the same display position but in temporal order (2IFC). For a given stimulus exposure time one can gather twice as many responses with 2AFC compared to 2IFC for the same session duration, but one must be careful. If the intention is to present the stimuli to parafoveal or peripheral vision, 2AFC is the preferred method because when the stimuli are placed either side of fixation the observer is less inclined to make an eye movement to one or the other stimulus and unintentionally foveate it, a temptation that is harder to resist with 2IFC. However, if the intention is that both stimuli be scanned foveally, then the 2IFC version is preferable. If the presentation time of a 2AFC task is too short (<1 second) observers may become frustrated while attempting to scan both stimuli within the time allocated. If presentation time is too long, the time-advantage of 2AFC is lost, and one might as well use 2IFC. Typically, proportion correct is used as the measure of performance with 2AFC/2IFC, but observers can sometimes be biased towards responding to one location/interval more than the other, in which case proportion correct is not a good measure and d' should be used (Chapter 6).

1AFC Same-different

In this task observers are presented with a pair of stimuli on each trial, with half the trials containing a pair that is the same and half the trials a pair that is different. The task is to decide whether the pair on each trial is the "same" or "different" (Figure 3.2). The main reason for using a same-different task is that the observer does not need to know the basis on which the discriminands differ. This is desirable in a variety of situations. One situation is when the experimenter is not sure on what basis observers will discriminate the stimuli, for example if the stimuli are faces with different expressions, and is loath to give precise instructions as to what observers should look for. Another situation is when the experimenter wants to present observers with multiple discriminand pairs from across a wide range of a stimulus dimension. For example, one might want to obtain an overall measure of orientation discrimination across all orientations, or an overall measure of color discrimination across a variety of different colors. In these circumstances it is preferable not to burdon observers with having to learn the basis for discriminating each stimulus pair; this can be especially difficult with circular dimensions such as orientation or color.

In the 1AFC version of the same-different task only two stimuli are involved, say S_1 and S_2. Hence there are two Same combinations, S_1S_1 and S_2S_2, and two Different combinations, S_1S_2 and S_2S_1. All four combinations are typically presented an equal number of times or with equal probability during a session. Because the two discriminands (Same and Different) are not symmetric, this task is particularly prone to the effects of bias, in this case a tendency towards responding "same" or towards responding "different." Thus, it is advisable to analyze the data to take into account any bias (see Chapter 6). The less-bias-prone 2AFC version of same-different is described later. The 1AFC same-different task is popular in animal experiments as a method for determining an animal's ability to recognize a previously-shown object. When employed for this purpose, the two stimuli are presented in temporal order and the animal is typically rewarded after correctly identifying a Same stimulus (e.g., Vallentin & Nieder 2008).

3.2.1.1.3 $N = 3$

3AFC Oddity

In the oddity task, sometimes termed "odd-man-out," all stimuli bar one are the same and the observer selects the stimulus that is different (Figure 3.2). Like the same-different task, an attractive feature of the oddity task is that the observer need not know the basis on which the stimuli differ. The minimum N for an oddity task is 3, and this version, sometimes termed the "triangular method," is undoubtedly the most popular (e.g., Huang, Kingdom, & Hess, 2006). Oddity tasks can be either three-alternative (3AFC) or three-interval (3IFC). With the 3AFC version the three stimuli are best positioned in a triangular arrangement on the screen (e.g., Pitchford & Mullen, 2005).

Are there disadvantages to the oddity task? Some observers find it difficult and frustrating. In the case of the 3IFC version, for example, the observer needs to hold in short-term memory three pieces of information prior to making a decision, and observers sometimes report difficulty remembering "what the first stimulus looked like." The 3AFC version avoids this problem providing observers are given plenty of time to compare all three stimuli, and probably the most successful version of the oddity task is the 3AFC version with unlimited stimulus exposure. However, many experimenters prefer the 2AFC match-to-sample or 2AFC same-different task to the 3AFC oddity task, for reasons now discussed.

2AFC Match-to-sample

In this task the observer views a "sample" stimulus and is then required to select the same stimulus from one of two "match" stimuli. As with the oddity and same-different tasks, the observer does not need to know the basis on which the stimuli differ. Match-to-sample tasks are particularly popular in animal (e.g., Jordan, MacLean, & Brannon, 2008), child vision (Pitchford & Mullen, 2005), and cognitive vision studies, such as on face recognition (e.g., Wilbraham et al., 2008). A particularly attractive feature of the match-to-sample task is that it can be used to study recognition memory, since the time delay between sample and match can be a variable. Part of the reason for the task's popularity is that it is an easy task for human observers to understand and for animals to learn. The relative ease with which it can be understood and learned may in part be due to the fact that the "same as" concept is easier to grasp than the "different from" concept needed for both the oddity and same-different tasks. The match-to-sample task is also less cognitively demanding than the oddity task, because there is one less alternative to choose from.

3.2.1.1.4 $N = 4$
2AFC/2IFC Same-Different

In this form of same-different task, the two pairs of stimuli, Same and Different, are presented together on a trial, and the observer chooses the pair that is Different (or Same). This version of same-different is less prone to bias than the 1AFC version described above, and for this reason is preferable. Because there are four stimuli per trial, a popular scenario is to present the two members of each pair together on the display, but in temporal order (e.g., Yoonessi & Kingdom, 2008). Presenting all four stimuli one after the other will likely be too cognitively demanding. Observers often prefer the 2AFC same-different task to the 3AFC oddity task because, although the former involves one extra stimulus, there is one less alternative to have to choose from on each trial.

3.2.1.1.5 $N > 4$
M-AFC tasks

Although we have argued that small-*N* forced-choice procedures are generally preferable to large-*N* ones, there are experimental questions that demand large *N*, and for this the standard forced-choice, oddity, and match-to-sample tasks are all available. The *M*-AFC match-to-sample task in particular is a very flexible tool offering a myriad of design possibilities. One can use the task for testing the ability of observers to select a sample not only from two stimulus states, but also from a large number of stimulus states, for example a red object from an array of green, red, yellow, blue, etc., match objects. Moreover, the stimuli can be defined along multiple dimensions, such as color and form. For example the observer might be required to select a red T-shape from an array of green O-, yellow B-, red T-, Blue Z-, etc., shapes. Another variant is to require observers to select the match that has one attribute in common with the sample even though differing in all other attributes, for example to select the match with the same color as the sample, even though different in form and motion, or to select the match with the same motion as the sample, even though different in form and color (e.g., Pitchford & Mullen, 2001).

3.2.1.2 Non-forced-choice Thresholds

3.2.1.2.1 Method of Adjustment

The method of adjustment is rarely used nowadays to obtain performance measures, since forced-choice procedures are easy to set up on a computer and are widely regarded as superior. However, the method of adjustment can be useful for obtaining a rough threshold estimate in order to guide the choice of stimulus magnitudes for an in-depth experiment, especially when there are a large number of different conditions to be measured (e.g., Nishida, Ledgeway, & Edwards, 1997).

3.2.2 Non-threshold Tasks Procedures

3.2.2.1 Accuracies and Reaction Times

Accuracy refers to how close a measure is to its true value, and can be measured using both forced-choice and method-of-adjustment. Examples of accuracy measures are described in the previous chapter.

Reaction times refer to the time taken for an observer to respond to the onset or offset of a stimulus. Reaction time is an important aptitude measure, and is often an accompaniment to other performance measures such as proportion correct (e.g., Ratcliff & Rouder, 2009). Reaction times are used as the main performance measure in studies of visual search, where the experimenter is interested in the time taken by observers to find a target among a set of distractors (e.g., Treisman & Gelade, 1980; McIlhagga, 2008). The analysis of reaction time data and its value for understanding psychological processes is a large topic that is outside the scope of this book; some useful summaries of the field are given at the end of the chapter.

3.3 APPEARANCE-BASED PROCEDURES

All appearance-based procedures are Type 2, since there can never be a correct and an incorrect response to a judgement about appearance. We have chosen to divide appearance procedures into matching and scaling, and then subdivide each of these categories into forced-choice and non-forced-choice. Note, however, that as we said in the previous chapter, matching and scaling procedures constitute only a fraction of the procedures used to measure stimulus appearance. Matching procedures aim to measure the point of subjective equality (PSE) between two stimuli. Although matching procedures can be used to derive perceptual scales, scaling procedures are explicitly aimed at uncovering the relationship between the perceived and physical magnitudes of a stimulus dimension. Let us first consider matching.

3.3.1 Matching

In Chapter 2 we described a number of matching experiments. The Rayleigh match aimed to determine which combinations of wavelengths matched a single, narrowband wavelength in both brightness and hue. The brightness-matching experiment measured the luminance of a disc that matched the brightness of a test disc surrounded by an annulus. The Muller–Lyer illusion experiment measured the length of the center line of one of the two Muller–Lyer figures that matched the perceived length of the center line of the other. The vernier experiment measured the offset of two lines at which they appeared aligned. Each of these experiments measured some form of the point of subjective equality (PSE) between physically different stimuli. Although the term "matching" conjures up an image of an observer adjusting something until it looks like something else, three of the above experiments used forced-choice procedures rather than the method of adjustment. With a forced-choice matching procedure the observer is required to make a comparative judgement of two stimuli on each trial, for example which stimulus looks brighter, which stimulus looks longer, etc., but the goal is in every case to establish a PSE.

3.3.1.1 *Forced-choice Matching*

3.3.1.1.1 $N = 2$: Matching Using 2AFC/2IFC

The reader is once again referred to the examples of the brightness-matching, Muller–Lyer, and vernier acuity experiments described in Chapter 2. There is little to add here except to emphasize that a forced-choice procedure enables the experimenter to derive a full psychometric function, and thus to obtain estimates of parameters beside the PSE and precision, such as the errors on the parameters. Full details of how to obtain parameter estimates from appearance-based psychometric functions are provided in Chapter 4.

3.3.1.2 *Non-forced-choice Matching*

3.3.1.2.1 $N = 2$: Matching Using Adjustment

Adjustment is still widely employed to obtain PSEs. Observers freely adjust one stimulus, termed the "match," "adjustable," or "variable" stimulus, until it appears equal along the dimension of interest to the "test" stimulus. If enough matches are made the variance or standard deviation of the settings can be used to provide a measure of precision.

3.3.1.2.2 $N = 2$: Nulling Using Adjustment

A variant on matching that frequently uses the method of adjustment is "nulling" or "cancellation." In some instances nulling and matching can be considered two sides of the same coin. Consider, for example, the brightness-matching experiment illustrated in Figure 2.5 (previous chapter). One can think of the annulus as inducing an "illusory" brightness in the test patch, because even though the luminance of the test patch remains fixed, its brightness changes with the luminance of the annulus. However, instead of the observer adjusting the luminance of the match patch to match the brightness of the test patch for each annulus luminance, the observer

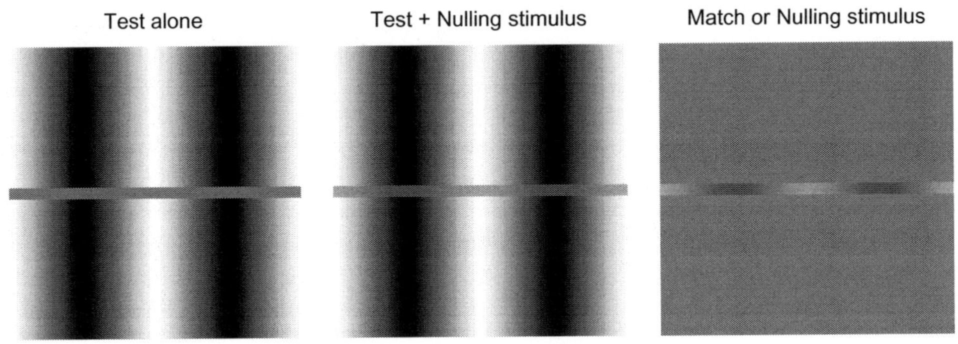

FIGURE 3.3 Matching versus nulling. Left: grating induction stimulus. The horizontal gray stripe running through the middle of the luminance grating is uniform yet appears modulated in brightness due to simultaneous brightness contrast. Right: an adjustable second grating with similar spatial dimensions to the induced grating can be used to match its apparent contrast. The same grating, however, can also be used instead to null or cancel the induced grating when added to it, as in the middle figure. Note that the cancellation is not perfect, because of the limitations of reproduction. See text for further details.

could instead "null" or "cancel" the effect of annulus luminance by adjusting the *test* luminance to match that of the fixed-in-luminance match patch. By the same token, if the observer adjusted the length of the central line of one of the Muller–Lyer figures (say the one with acute fins) until it matched that of the length of the line in the other Muller–Lyer figure (the one with obtuse fins), one could say that the illusion was being nulled or cancelled.

The difference between nulling and matching emerges more forcefully when applied to the grating-induction illusion illustrated in Figure 3.3 (McCourt, 1982). In the left figure one observes an illusory modulation in brightness in the gray stripe that runs horizontally through the grating. The modulation is illusory because the gray stripe is actually uniform in luminance. The illusory modulation is an instance of the well-known phenomenon termed "simultaneous brightness contrast." Notice how the illusory brightness modulation is out-of-phase with the real luminance modulation in the surround grating (i.e. the ordering of bright and dark is opposite). The strength or contrast of the illusory or "induced" modulation depends on a number of factors, and to study these factors one needs a method of measuring the size of the induction. Two methods are illustrated in Figure 3.3. The matching procedure uses a second grating with similar spatial dimensions to the induced grating, as illustrated on the far right of the figure. The observer adjusts the contrast of the matching grating until it appears equal in apparent contrast to that of the induced grating. The contrast of the matching grating is typically measured using the metric of contrast known as Michelson contrast, defined as $(L_{max} - L_{min})/(L_{max} + L_{min})$, where L_{max} and L_{min} are the maximum and minimum luminances of the grating. Thus, with the matching procedure, the magnitude of brightness induction is measured by the contrast of the matching grating at the PSE. In the nulling procedure, on the other hand, the second grating is *added* to the induced grating and its contrast adjusted until the induced grating just disappears (McCourt & Blakeslee, 1994) – this is illustrated in the middle figure. Note that with the nulling procedure the phase of the added grating must be opposite to that of the induced grating in order for the cancellation to work, as in the figure (this is not necessary for the matching procedure). With the nulling procedure, the contrast of the nulling grating that cancels the induced grating is the measure of the size of the induction.

3.3.2 Scaling

3.3.2.1 *Types of Perceptual Scale*

Recall that a perceptual scale describes the relationship between the perceptual and physical magnitudes of a stimulus dimension. There are three types of perceptual scale that are most relevant to psychophysics: ordinal; interval; and ratio. In an ordinal perceptual scale, stimulus magnitudes are numbered according to their rank order along the perceptual continuum. However, the difference between any pair of numbers does not necessarily correspond to the magnitude of the perceptual difference. For example, consider a stimulus with three contrasts: 0.1; 0.7; and 0.8. On an ordinal scale these might be numbered 1, 2, and 3, but this does not imply that the perceptual difference between the 0.1 and 0.7 contrasts on the one hand, and between the 0.7 and 0.8 contrasts on the other, are equal. On the contrary, the perceptual differences will almost certainly be very different. To represent the perceptual differences between these pairs of contrasts, an interval or ratio scale is required. In an interval scale, the differences between numbers correspond to perceptual differences, even though the numbers themselves are arbitrary. Using the example of the three contrasts above, an interval scale might be 1, 5, and 6. This time the numbers capture the observation that the perceptual difference between the first and second contrasts – a difference of four scale units – is four times greater than the perceptual difference between the second and third contrasts – a difference of one scale unit. However, the interval scale could just as easily be written 4, 12, and 14, since these numbers embody the same difference-relations as the 1, 5, and 6 scale. Formally, an interval scale can be transformed without loss of information by the equation $aX + b$, where X is the scale value, and a and b are constants.

The limitation of an interval scale is that it does not capture the perceived relative magnitudes of the stimulus dimension. For example, interval scale values of 1 and 5 do not indicate that the second value is five times the perceived magnitude of the first. Perceptual scales that capture relative perceived magnitude are known as ratio scales, and can be transformed only by the factor aX.

The relationship between perceived and physical contrast is an example of a one-dimensional perceptual scale. However, perceptual scales can be two-dimensional.

The best-known example of a two-dimensional perceptual scale is a color space (such as the CIE), in which each color is defined by a point on a plane with an X and a Y coordinate, and where the distance between points corresponds to the perceived distance in hue, or perceived chromaticity. Two-dimensional perceptual scales are invariably interval scales. Figure 3.4 summarizes the main varieties of one-dimensional interval-scaling tasks that are now described.

3.3.2.2 *Forced-choice Scaling Procedures*

3.3.2.2.1 $N = 2$: Paired Comparisons

The simplest forced-choice method for deriving a perceptual scale is the method of paired comparisons. If the stimulus space is sampled only coarsely, paired comparisons can only provide an ordinal perceptual scale. For example, suppose one wants to rank order, say, ten photographs of faces according to how happy they look. On each trial observers are shown two faces drawn from the set and asked to indicate which face looks happier. There would be a total of $(10^2 - 10)/2 = 45$ possible face pairs, or twice this number if every pair was shown in both order. On each trial the face selected to be the happier is given a score of one, while the other face is given a score of zero. If the procedure is repeated for all possible pairs of faces, the ten faces can be rank-ordered by perceived happiness according to their accumulated scores.

In order to generate an interval scale using paired comparisons, however, the different stimulus levels must be close enough to ensure that the responses to any pair are not always the same. Instead, the data must be a "proportion" of times that one member of a pair is chosen over the other. With proportions as the data one can estimate the perceptual distances between stimulus levels, and hence generate an interval perceptual scale. Chapter 7 describes the Palamedes routines for generating stimulus lists, simulating observer responses, and analyzing responses to produce an interval perceptual scale using the method of paired comparisons. The chapter also includes a critical discussion of the strengths and limitations of the paired comparison method.

3.3.2.2.2 $N = 3$: Method of Triads

This method can also be used to derive either an ordinal or interval scale, but uses judgements of relative perceived similarity (or difference). Unlike the $N = 2$ paired-comparison method, the method of triads does not require prior knowledge of the dimension along which the stimuli differ.

In one version of the method of triads, one of the three stimuli is designated the target, the other two the comparisons. The observer is required to compare the perceived similarity (or difference) between the target and each of the two comparisons, and choose the pair that is the more (or less) similar. Typically, the pair

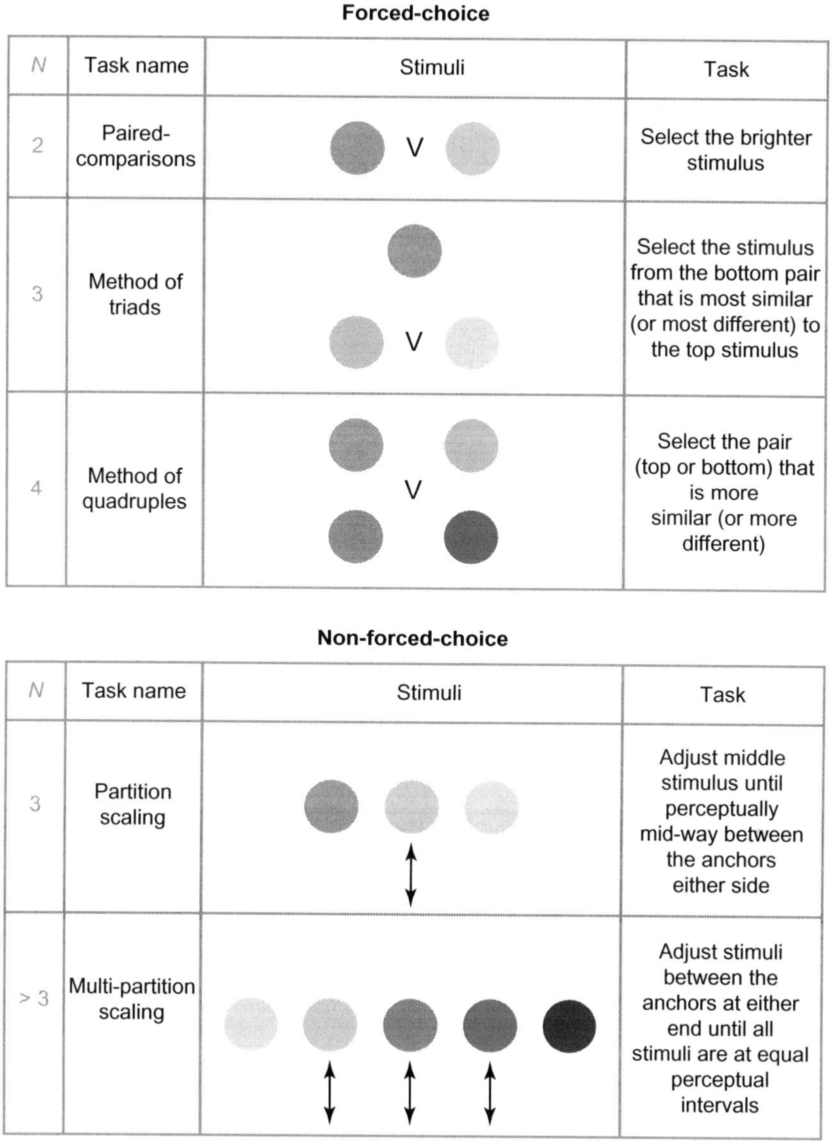

FIGURE 3.4 Types of scaling task for deriving interval scales, applied to the example of brightness scaling. In the non-forced-choice methods in the lower panel the double arrows refer to disks whose luminances are freely adjusted by the observer.

perceived to be more similar would be given a value of one, the pair perceived to be less similar a value of zero. One can think of this version of the task as the appearance analog of the 2AFC match-to-sample performance task described earlier in the chapter. In another version of the method of triads there is no designated target, and the observer compares all of the three possible pairs, giving ranking them one, two, or three. Palamedes routines for deriving an interval perceptual scale using the method of triads is described in Chapter 7.

3.3.2.2.3 $N = 4$: Method of Quadruples

In this procedure, observers are presented with two pairs of stimuli on each trial, and the task is to decide which pair is the more (or less) similar. As with the method of triads, the observer need not know the basis on which the stimuli differ. The Palamedes routines for deriving interval scales using the method of quadruples are described in Chapter 7.

3.3.2.2.4 $N > 4$: Multi-stimulus Scaling

An alternative to the paired comparison method for deriving an ordinal perceptual scale is to present observers with the entire stimulus set together and ask them to arrange the stimuli in rank order. The best known example of this method is the Farnsworth–Munsell 100 hue test for color deficiency. Observers are presented with a randomly-arranged series of disks that vary systematically along a particular color dimension (e.g., green to red), and are asked to arrange them in order according to hue (e.g., green, yellowish-green, more-yellowish-green, yellow, reddish-yellow, more-reddish-yellow … red). The resulting arrangement is compared to that typically made by a person with normal color vision. One can think of the order made by a person with normal color vision as the "correct" order, but it is only "correct" in relation to an internal standard, not to a physical standard as with a Type 1 experiment. The pattern of errors made by observers with the Farnsworth–Munsell test can be used to identify certain types of color deficiency.

3.3.2.2.5 Multi-dimensional Scaling

Multi-dimensional scaling (MDS) is used to determine whether two or more perceptual dimensions underlie the perceived similarities between stimuli. Earlier we mentioned the CIE color space as an example of a two-dimensional representation of perceived color similarities. MDS algorithms provide multi-dimensional arrangements of stimuli in which the distances between stimuli correlate with their perceived dissimilarity. The method of triads and quadruples can be used to generate data for MDS (e.g., Gurnsey & Fleet, 2001). The analysis of MDS data is, however, outside of the scope of this book, but some example reading material is provided at the end of this chapter.

3.3.2.3 *Non-forced-choice Scaling Procedures*

3.3.2.3.1 $N = 1$: Magnitude Estimation

In magnitude estimation the observer makes a direct numerical estimate of the perceived magnitude of the stimulus along the dimension of interest. Magnitude estimation produces a ratio scale if observers are instructed to allocate numbers that reflect the relative perceived magnitudes of the stimuli. In one form of magnitude estimation, the experimenter starts with a stimulus designated as an "anchor" and asks the observer to suppose that it has a perceived magnitude of, say, 50. The other stimuli are then estimated relative to the anchor, i.e., 25 (half as much), 100 (twice as much), 175 (3.5 times as much), etc. The scale values can then be normalized to the stimulus with lowest perceived magnitude by dividing all values by 50. Psychophysicists tend to regard magnitude estimation as a rather blunt tool, because it requires observers to translate a perceptual experience into a numeric, i.e., symbolic, representation. Observers often find magnitude estimation difficult and unsatisfactory, and for this reason other scaling methods are recommended whenever possible.

3.3.2.3.2 $N = 3$: Partition Scaling

In partition scaling, sometimes termed "equisection" or "bisection" scaling, observers adjust the magnitudes of stimuli in order to make them appear at equal perceptual intervals. Partition scaling methods therefore generate interval scales. There are a variety of partition scaling methods, and the principle behind two of them is illustrated at the bottom of Figure 3.4. One version that is intuitively easy for the observer, but which has some drawbacks, is termed by Gescheider (1997) the "progressive solution." The experimenter starts by providing the observer with two "anchors" that define the start and end points of the stimulus dimension. The observer then divides the perceptual distance between the two anchors into two equal parts by adjusting a third stimulus until it appears perceptually midway between the anchors.[1] The resulting two intervals are then each bisected in a similar manner, resulting in four intervals, and so on. This method, however, suffers from the problem that errors will tend to accumulate as the intervals become smaller.

3.3.2.3.3 $N > 3$: Multi-partition Scaling

In what Gescheider (1997) terms the "simultaneous solution," and termed here multi-partition scaling (Figure 3.4), observers are presented with the full set of stimuli together on the display. Two stimuli at the ends of the range serve as anchors, and observers adjust the remaining stimuli until they appear to be at equal perceptual intervals. Recall Whittle's (1992) multi-partition scaling experiment described in

[1]Note that the bisection scaling task is different from the bisection acuity task described in Chapter 2. The latter is a performance-based task that measures the accuracy and/or precision of bisecting a line.

Chapter 2. The aim of the experiment was to derive an interval scale of brightness for discs of adjustable luminance arranged in the form of a spiral on a uniform gray background. The anchor discs were set to the lowest and highest luminances available on the monitor and observers adjusted the luminances of the remaining discs until they appeared to be at equal intervals in brightness. Intuitively, this is not an easy task, since adjustment to any one disc would tend to "throw out" previous adjustments, requiring a number of iterations to achieve a perceptually satisfactory solution.

3.4 FURTHER DESIGN DETAILS

3.4.1 Method of Constant Stimuli

In any forced-choice procedure, whether performance-based or appearance-based, the question arises as to how to present the different magnitudes of a stimulus during an experimental session. One popular solution is the method of constant stimuli, or as it is sometimes termed, the "method of constants." In this method, the stimulus magnitude on each trial is randomly selected from a predefined set. For a performance-based experiment, the range is typically chosen to straddle the expected threshold value in order that performance ranges from near-chance to near-100% correct. For example, in a standard 2AFC procedure with threshold defined at the 75% correct level, performance should range from close to 50%, to close to 100%, with roughly equal numbers of stimulus magnitudes producing less than and greater than 75% correct. This generates data that, when fitted with the appropriate psychometric function, provides the most accurate estimates of the threshold as well as other parameters, such as the slope. Full details of the procedures for fitting psychometric functions are described in Chapter 4. The choice of stimulus set usually requires some pilot work to obtain a rough estimate of the threshold, and the method of adjustment is useful for doing this.

The method of constant stimuli can also be used in conjunction with appearance-based procedures. For forced-choice matching experiments in which the PSEs are estimated from a psychometric function, the above considerations equally apply, though this time the data are not proportions correct but proportions of times one stimulus is perceived to be greater than the other along the dimension of interest.

3.4.2 Adaptive Procedures

To avoid the problem of inappropriately-chosen stimulus sets, adaptive (or staircase) procedures are often used instead of the method of constant stimuli. In an adaptive procedure the stimulus magnitude on each trial is selected by an algorithm that analyzes the previous trial responses, in order to "zero in" on the threshold.

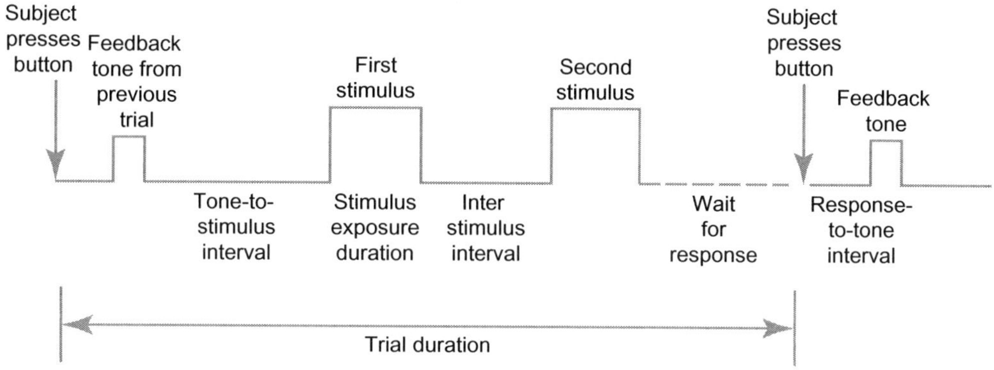

FIGURE 3.5 Example timing of stimulus presentation during a typical 2IFC trial.

Some adaptive procedures can be used in conjunction with conventional methods for fitting psychometric functions, enabling estimates of both the threshold and slope to be obtained. Adaptive methods can be used in conjunction with both performance-based and appearance-based tasks. Adaptive procedures are the subject of Chapter 5.

3.4.3 Timing of Stimulus Presentation

The timing of stimulus presentations is very important in psychophysics, and for an observer can make the difference between an experiment that feels difficult and frustrating and one feeling comfortable and engaging. To illustrate what's at stake, take a prototypical 2IFC task. Figure 3.5 is a schematic of the temporal arrangement and terminology. The example is of an observer-paced trial, in which each trial is triggered by the observer's response to the previous trial. In general, 2IFC tasks are best when self-paced, as this gives the observer control over the pace of the experiment, without disrupting the critical temporal parameters.

The choice of within-trial temporal parameters is crucial for making a task feel comfortable. For example, if the first stimulus of the forced-choice pair is presented too soon after the observer responds to the previous forced-choice pair, the observer can become confused as to what his/her response is "attached to;" the response may become associated in the observer's mind with the stimulus that follows it rather than with the stimulus that precedes it. An appropriate inter-stimulus interval (ISI) is also important to minimize both forward and backward masking effects between stimuli. There is no hard-and-fast rule here, and the experimenter needs to try out different ISIs until the task feels comfortable. As a rule of thumb, a stimulus exposure duration of 250 ms, an ISI of 500 ms, and an inter-trial-interval (ITI) of 1000 ms is a good starting point.

Further Reading

An excellent and user-friendly introduction to much of what is discussed here can be found in Gescheider (1997). Reviews of the use of reaction times in psychological research can be found in Pachella (1974) and Meyer et al., (1988). Multidimensional scaling is discussed in Borg & Groenen (2005).

References

Borg, I., & Groenen, P. J. F. (2005). *Modern Multidimensional Scaling: Theory and Applications*. New York: Springer.

Gurnsey, R., & Fleet, D. J. (2001). Texture space. *Vision Research, 41*, 745–757.

Huang, P.-C., Kingdom, F. A. A., & Hess, R. F. (2006). Only two phase mechanisms, ± cosine, in human vision. *Vision Research, 46*, 2069–2081.

Jordan, K. E., MacLean, E., & Brannon, E. M. (2008). Monkeys match and tally quantities across senses. *Cognition, 108*, 617–625.

McCourt, M. E. (1982). A spatial frequency dependent grating-induction effect. *Vision Research, 22*, 119–134.

McCourt, M. E., & Blakeslee, B. (1994). A contrast matching analysis of grating induction and suprathreshold contrast perception. *Journal of the Optical Society of America A, 11*, 14–24.

McIlhagga, W. (2008). Serial correlations and 1/f power spectra in visual search reaction times 5, 1–14. *Journal of Vision, 8*(9).

Meyer, D. E., Osman, A. M., Irwin, D. E., & Yantis, S. (1988). Modern mental chronometry. *Biological Psychology, 26*, 3–67.

Nishida, S., Ledgeway, T., & Edwards, M. (1997). Dual multiple-scale processing for motion in the human visual system. *Vision Research, 37*, 2685–2698.

Pachella, R. G. (1974). The interpretation of reaction time in information-processing research. In B. H. Kantowitz (Ed.), *Human Information Processing: Tutorials in Performance and Cognition* (pp. 41–82). Hillsdale, N.J.: Lawrence Erlbaum Associates.

Pitchford, N. J., & Mullen, K. T. (2001). Conceptualization of perceptual attributes: a special case for color? *J. Experimental Child Psychology, 80*, 289–314.

Pitchford, N. J., & Mullen, K. T. (2005). The role of perception, language, and preference in the developmental acquisition of basic color terms. *J. Experimental Child Psychology, 90*, 275–302.

Ratcliff, R., & Rouder, J. N. (2009). Modeling response times for two-choice decisions. *Psychological Science, 9*, 347–356.

Treisman, A. M., & Gelade, G. (1980). A feature-integration theory of attention. *Cognitive Psychology, 12*, 97–136.

Vallentin, D., & Nieder, A. (2008). Behavioral and prefrontal representation of spatial properties of monkeys. *Current Biology, 18*, 1420–1425.

Wilbraham, D. A., Christensen, J. C., Martinez, A. M., & Todd, J. T. (2008). Can low level image differences account for the ability of human observers to discriminate facial identity? 5, 1–12. *Journal of Vision, 8*(15).

Yoonessi, A., & Kingdom, F. A. A. (2008). Comparison of sensitivity to color changes in natural and phase-scrambled scenes. *Journal of the Optical Society of America A, 25*, 676–684.

CHAPTER

4

Psychometric Functions

4.1 INTRODUCTION

Psychometric functions, or PFs, relate the behavior on a given psychophysical task (e.g., proportion of correct responses, proportion of trials perceived brighter) to some physical characteristic of the stimulus (e.g., contrast, length). Typically, although not always, one measures a PF in order to determine one or more parameters that summarize the behavior, e.g., a threshold contrast or a point of subjective equality. Chapter 2 showed examples of PFs and the parameters determined by them. In this chapter we will introduce the reader to methods of determining the chosen parameters from a PF. It is important to note that PFs can be fitted to the data for both performance-based and appearance-based psychophysical tasks, and by and large the procedures for fitting PFs are common to both classes of data.

The chapter is divided into two sections. The first section introduces the reader to the general procedures involved in fitting PFs, determining the chosen parameters from the fits and getting estimates of how well the PFs have been fit, as well as the variability of the estimated parameters. The second section of the chapter will consider the underlying theory behind PFs, fitting PFs and parameter estimation. The reader may choose to read the second section or to skip it without loss of continuity.

4.2 SECTION A: PRACTICE

4.2.1 Overview of the Psychometric Function

Figure 4.1 illustrates the general idea. The figure shows hypothetical data from an experiment aimed at measuring a contrast detection threshold. The data were obtained from a 2IFC task using the method of constant stimuli. The graph plots the proportion of correct responses for each stimulus contrast. Note that the contrast values on the abscissa are arranged at equal logarithmic intervals. However, other arrangements such as linear spacing of values are often used. The observer performed 50 trials for each stimulus contrast. Threshold contrast is defined as the contrast at which the proportion correct response reaches some criterion, here 0.75 or 75%. In order to obtain the value corresponding to the threshold a continuous function has been fitted to the data. The function in this example is known as

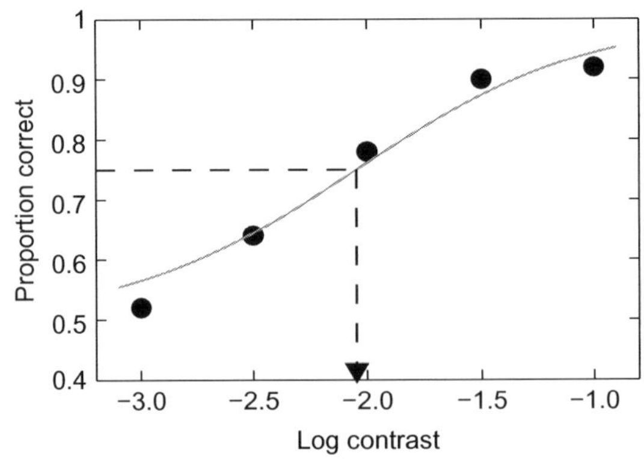

FIGURE 4.1 Example of a psychometric function from a hypothetical experiment aimed at measuring a contrast detection threshold. The threshold is defined here as the stimulus contrast at which performance reaches a proportion correct equal to 0.75. Data are fitted using a Logistic function.

a Logistic function, and is one of a variety of functions that can be used to fit PFs. To fit the logistic curve to the data the computer iteratively searched through a range of possible values of two parameters, α (alpha) and β (beta). The α parameter determines the overall position of the curve along the abscissa, and for the Logistic function corresponds to the contrast value at which the proportion correct is halfway between the lower and upper asymptote, here 0.75 or 75% correct. The β parameter determines the slope or gradient of the curve. The parameters α and β are properties of the observer and we will never know their exact value. Rather, the fitting procedure found estimates of the values of α and β that generated a curve that best matched the experimental data. We use a "hat" over the symbol for a parameter to mean "estimate of" that parameter. Thus, the value of $\hat{\alpha}$ for the best-fitting curve is the estimate of the true contrast threshold, α, and $\hat{\beta}$ is the estimate of the true slope β.

Four additional values make up the complete description of the PF. Two of these, the "standard error" (SE) of the threshold and the SE of the slope, are measures of the precision, or rather imprecision, of $\hat{\alpha}$ and $\hat{\beta}$, i.e., how far they are likely to be from the "true" value of α and β. Put another way, they are estimates of the errors associated with our estimates of α and β. The remaining two measures, "deviance" and its associated p-value, are used to determine whether the fitted function provides an adequate model of the data. We will discuss goodness-of-fit briefly in Section 4.2.6 and in much more detail in Chapter 8. Table 4.1 gives the values of all six of these measures for the PF in Figure 4.1.

This example PF illustrates the key components to measuring and fitting a psychometric function. In summary these are: (1) choosing the stimulus levels; (2) selecting the function to fit the data; (3) fitting the function; (4) estimating the errors on the function's parameter estimates; (5) determining the goodness-of-fit of the function. In what follows we consider these components in turn.

4.2.2 Number of Trials and Stimulus Levels

4.2.2.1 Number of Trials

How many trials are needed to estimate a psychometric function? As a rule, the more trials there are the more accurate the estimates on the fitted parameters, such

TABLE 4.1 Six values describing a fitted psychometric function

Threshold $\hat{\alpha}$	Slope $\hat{\beta}$	SE threshold	SE slope	Deviance	p value
−2.046	1.984	0.1594	.6516	1.35	0.738

as the threshold, slope, or point of subjective equality (PSE) will be. So the answer to the question primarily rests on how precise one wants one's parameter estimates to be. If one anticipates that different conditions will produce very different thresholds or PSEs, then this can be demonstrated in fewer trials than if one anticipates that there will be slight differences in thresholds or PSEs. In addition, curve fitting procedures might not converge on a fit if there are insufficient data. Although there is no hard-and-fast rule as to the minimum number of trials necessary, 400 trials is a reasonable number to aim for when one wants to estimate both the threshold and the slope of the PF.

4.2.2.2 *Range of Stimulus Levels*

As a general rule-of-thumb, for a performance-based task one wants to choose a set of stimulus levels that will result in performance that ranges from just above chance to just under 100% correct. If more than one stimulus level produces approximately chance performance this means that the lower end of the stimulus range needs to be shifted to a higher level. Similarly, if more than one stimulus level produces approximately 100% correct performance, the highest stimulus level needs to be shifted to a lower level. There is no need to use many, finely spaced stimulus levels. Concentrating responses at just a few appropriately distributed stimulus levels should suffice to obtain reliable estimates of the parameters of a PF.

We will have much more to say about the number of trials needed, as well as the range of stimulus values to use, in Chapter 5. Chapter 5 will discuss adaptive testing methods which, in essence, aim to increase the efficiency of the experimental procedure. That is, they aim to gather the greatest amount of information as possible about the PFs parameters of interest while using as few trials as possible. They do this by presenting the stimuli at levels that are expected to provide the most information possible about the parameters of interest.

4.2.2.3 *Linear Versus Logarithmic Spacing of Stimulus Levels*

An issue that always seems to come up is how to space the values of the independent variable. The two choices are usually linear or logarithmic (log). Which one should one choose? One of the reasons given for log spacing is that it allows for a greater range of values. However, this makes no sense – you can have as big a range with linear spacing as with log spacing. Whether using linear or log spacing, the bigger the range, the bigger the interval between values. A more sensible reason for using logarithmic spacing is that you want relatively small intervals at the low and relatively large intervals at the high end of the range. One reason for wanting the interval to increase with the stimulus value is because this gives a closer approximation to how intervals might be represented in the brain. The relationship between the physical and internal representation of a dimension is called the "transducer function." In the auditory and visual domains, these transducer functions are

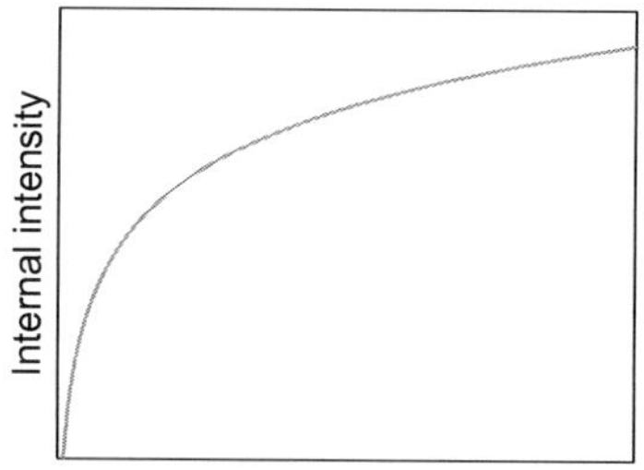

FIGURE 4.2 Typical transducer function between the physical intensity of a stimulus and the corresponding subjective or internal intensity.

generally decelerating, such as that shown in Figure 4.2. That is, as stimulus intensity increases constant increases in stimulus intensity lead to smaller and smaller increases in internal intensity. Of course, the precise form of the bow-shape varies between dimensions, and if one knew the shape exactly, one could space the corresponding x-axis value accordingly. But, given that most dimensions have a bow-shaped transducer function, log spacing of values is a "good bet."

To derive a set of logarithmically-spaced values you need to decide on the first and last values of the series (call these a and b), and how many values you want (call this n). The ith value of the sequence ($i = 1 \ldots n$) is given by the equation:

$$x_i = 10^{[\log a + (i-1)\log(b/a)/(n-1)]} \tag{4.1}$$

The MATLAB® function **logspace** implements this equation:

```
>>StimLevels = logspace(log10(a),log10(b),n)
```

For example, suppose you want five values of contrast, ranging from 0.025 to 0.8. Thus $a = 0.025$, $b = 0.8$ and $n = 5$. Substituting these values in the above command will output:

```
StimLevels = 0.0250 0.0595 0.1414 0.3364 0.8000
```

Note that the values are not actual log values, just values that are logarithmically spaced. The sequence is sometimes known as a geometric series, because the ratio of any pair of adjacent values is the same – you can check this yourself (note that the ratios will not be exactly the same because the numbers are only given with a maximum of four decimal places). If you choose to use log spacing, you have to be careful when fitting your psychometric function. If you are using a standard curve fitting program and enter the values above for the x-axis, the fitting procedure will not "take into account" the fact that your values are logarithmically spaced, and will just treat them as raw values. As a result the fit you get might not be particularly good, because the function you are fitting has a shape that may not correspond to the "stretching out" of values at the high end of the scale. The solution is to convert your x values into actual log values and *then* fit the psychometric function, because when you convert values that are logarithmically spaced into log values, they will be evenly spaced. Of course, having fitted your psychometric function to the actual log values you may want to report the raw thresholds. The raw threshold is simply the antilog of the log threshold.

4.2.3 Types and Choice of Function

4.2.3.1 Types of Function

In this section we introduce five functions that can be used to model psychometric data. They are: Cumulative Normal; Logistic; Weibull; Gumbel; and Hyperbolic Secant. Formal details of the equations of these functions, their strengths and limitations will be provided in Section B. To illustrate the shapes of these functions, Figure 4.3 shows an example of a set of data fitted with these functions. These functions all have the familiar sigmoidal shape. As can be seen, the estimate of the thresholds at the 0.75 correct level would be near-identical for all four functions.

In the Introduction we introduced two parameters that were estimated by the fitting procedure: α and β. These parameters describe properties of the underlying sensory mechanism. Two other parameters, however, are needed to specify the psychometric function fully. These are γ (gamma) and λ (lambda). These parameters do not correspond to properties of the underlying sensory mechanism, but rather describe chance-level performance and lapsing, respectively. We will discuss the two parameters in turn.

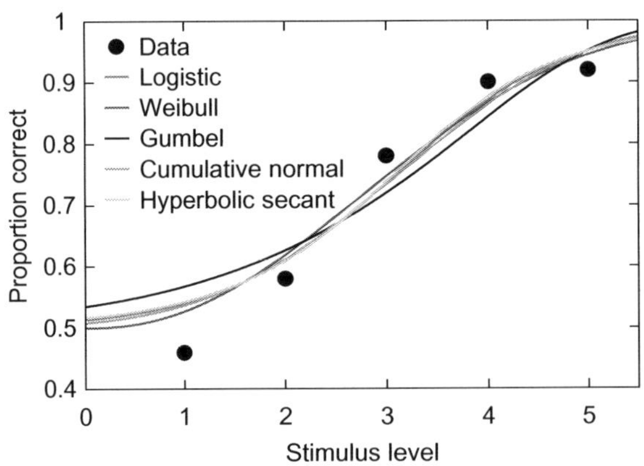

FIGURE 4.3 Example fits of five different PF functions. See separate color plate section for full color version of this figure.

The parameter γ is known as the guessing rate. In Section B we argue that this is a bit of a misnomer, since it is believed that an observer never truly guesses. Nevertheless, if an observer *were* to guess on a trial there is a certain probability that the guess would be correct. For example, in a performance-based task γ is simply assumed to equal the reciprocal of the number of alternatives in the forced-choice task, or $1/m$ in an M-AFC task (remember that m corresponds to the number of response choices in an M-AFC task. For all tasks described in this chapter $m = M$). Thus, for 2AFC γ is $1/2$ (0.5), for 3AFC it is $1/3$, etc. Figure 4.4 shows examples of the Logistic function fitted to performance-based 2AFC, 3AFC, and 4AFC data, with γ respectively set to 0.5, 0.33, and 0.25. Notice how the range of proportion correct on the y axis is different for the three plots. The proportion correct responses at the threshold parameter α vary with the guessing rate. Using the Logistic function, proportion correct at the threshold corresponds to 0.625, 0.667, and 0.75, respectively, for 4AFC, 3AFC, and 2AFC, respectively. For the Logistic function these values are calculated as $\gamma + (1 - \gamma)/2$.

The guess rate parameter is also important when fitting PFs for 2AFC data in which responses are coded not in terms of proportion correct but as proportions of one judgment over another, for example the porportion of "brighter" responses in a brighter-versus-darker brightness task, or the proportion of "left" responses in a left-versus-right vernier alignment task. In such experiments the resulting proportions range from 0-1, and the fitted PFs can be used to obtain either appearance-based measures such as PSEs (as in the brightness matching experiment), or performance-based measures such as accuracy (as in the vernier alignment experiment) and

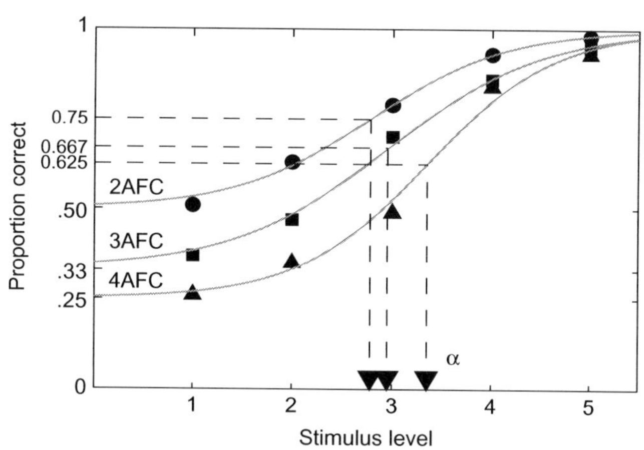

FIGURE 4.4 Example PFs for a 2AFC, 3AFC and 4AFC task. A Logistic function has been fit to each set of data by using a different value for the guessing parameter, γ (0.5, 0.33 and 0.25 respectively). The threshold α corresponds to proportion correct of 0.75, 0.667 and 0.625 respectively.

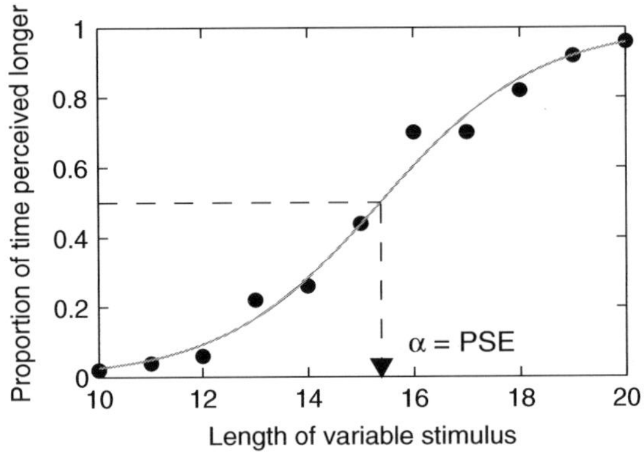

FIGURE 4.5 PF for an appearance-based task.

precision (as in both the brightness matching and vernier alignment experiments) (see Chapter 2 Section 2.3.4). Figure 4.5 shows data from a hypothetical appearance-based task. Suppose the aim of the experiment is to find the PSE for the length of two bars of different width. One bar is fixed in length, the other is varied. On each trial the task is to judge which bar appears longer. The length of the variable bar is shown on the abscissa. The ordinate of Figure 4.5 gives the proportion of times the variable bar is perceived as the longer. Thus, a Y value of 0.0 means that the variable bar was always perceived as shorter than the fixed bar, while a value of 1.0 means that the variable bar was always perceived as longer than the fixed bar. The PSE is the length of the variable bar which would be perceived just as many times shorter as it is perceived longer than the fixed bar. In other words, it would be perceived as longer on a proportion of 0.5 of the trials. To estimate the PSE parameterized by α, we have fitted a Logistic function and set the guessing rate γ to 0. Remember that, using the Logistic function, threshold α corresponds to a proportion correct in a performance-based task given by $\gamma + (1 - \gamma)/2$. For the appearance-based task we use the same equation, which gives 0.5 with $\gamma = 0$. This value of γ will be typical for appearance-based matching tasks. Some prefer to plot data in appearance-based tasks on graphs in which the ordinate ranges from -1 to $+1$, presumably so as to have a score of 0 correspond to the PSE, which has some intuitive appeal. This can simply be achieved by rescaling the plot in Figure 4.5. However, for the purposes of "fitting" the function, an ordinate scale of $0 - 1$ needs to be used. In the framework of theories of the psychometric function (Section 4.3.1 discusses two of these theories) the y-values correspond to the probabilities of observing one response rather than some other, and as such are constrained to have a value between 0 and 1, inclusive.

The fourth parameter associated with a PF, λ, is known as the lapse rate. On a small proportion of trials, observers will respond independently of stimulus level. For example, observers may have missed the presentation of the stimulus, perhaps due to a sneeze or a momentary lapse of attention. On such trials, observers may produce an incorrect response even if the stimulus level was so high that they would normally have produced a correct response. As a result of these lapses, the PF will asymptote to a value which is slightly less than 1. The upper asymptote of the PF corresponds to $1 - \lambda$. Note that if a lapse is defined as a trial on which the observer misses the presentation of a stimulus and consequently guesses, lapse rate is really not the appropriate term for the parameter λ. Rather, λ corresponds to the probability of responding incorrectly as a result of a lapse (but on some lapse trials, the observer will respond correctly by guessing). We will discuss this issue in more detail in Section B of this chapter.

Of the four parameters of the PF, researchers are typically interested only in threshold α and slope β, in that only α and β tell us something about the underlying sensory mechanism. The guessing rate γ is typically determined by the psychometric procedure (2AFC, 3AFC, etc.), and lapse rate λ tells us not about the sensory mechanism, but rather something about perhaps such things as the alertness or motivation of the observer. Researchers will usually allow only α and β to vary during the fitting procedure, and assume fixed values for γ and λ. Parameters that are allowed to vary during fitting are referred to as "free parameters," those that are not allowed to vary are referred to as "fixed parameters." The guessing rate in an *M*-AFC task can in most cases safely be assumed to equal $1/m$. However, it is debatable whether it is reasonable to assume any fixed value for the lapse rate. Researchers often implicitly assume the lapse rate to equal 0. Even the most experienced and vigilant observer, however, will occasionally respond independently of stimulus level. When it is assumed that lapse rate equals 0, but lapses do in fact occur, this may produce a significant bias on the threshold and slope parameters. The bias may be largely avoided if we allow a few lapses to occur by assuming the lapse rate to have a fixed small value, such as 0.01. An alternative, of course, is to make the lapse rate a free parameter, thereby simply estimating its value from the data. The issue will be discussed in more detail in Section B of this chapter. The function that allows one to find values of PFs in the Palamedes toolbox is of the general form:

```
y = PAL_[NameOfFunction](paramValues, x);
```

where **[NameOfFunction]** can be **CumulativeNormal**, **Logistic**, **Weibull**, **Gumbel**, or **HyperbolicSecant**. **paramValues** is a vector which contains values of the parameters of the PF (α, β, γ, λ), and **x** is a scalar, vector, or matrix containing the values at which the function should be evaluated. Try, for example, generating six values for the Logistic, first setting the stimulus levels to range from 1 through 6 as follows:

```
>>StimLevels = [1:1:6];
>>pcorrect = PAL_Logistic([3 1 0.5 0],StimLevels)
```

The output is:

```
pcorrect= 0.5596 0.6345 0.7500 0.8655 0.9404 0.9763
```

The command:

```
>>plot(StimLevels, pcorrect, 'ko');
```

will generate a crude plot of the PF.

The vector **paramsValues** does not need to contain values for all four parameters of the PF, but does need to contain at least the values for the threshold and the slope. If **paramsValues** contains only two entries, they are interpreted as values for the threshold and slope, respectively, and the guess rate and lapse rate are assumed to be zero. If a third value is provided it is interpreted as the guess rate, a fourth will be interpreted as the lapse rate. As an example, in the function call:

```
>>pcorrect = PAL_Logistic([3 1 0.5],StimLevels)
```

the vector passed to the function contains three values only, and as a result the lapse parameter is assumed to equal 0. The function thus returns the same results as above. Try generating **pcorrect** values using some of the other types of PF, and also investigate the effect of changing the four parameters: α, β, γ, and λ.

4.2.3.2 *Choice of Function*

What function should one choose? Essentially there are two criteria. The first is that one chooses the function based on an *a priori* theory of the "true" internal shape of the psychometric function. Different theories lead to the use of different functions, although the different functions that are in use are very similar (Figure 4.3), such that in practice the choice of function is often made based on convenience. In Section 4.3.2 we provide some of the theoretical background that might inform one as to which type of function one might want to choose, based on *a priori* considerations. The second criterion is based on *a posteriori* considerations, specifically using the function that most easily and accurately fits the data. Many practitioners, rightly or wrongly, base their choice on this second criterion.

Once a researcher has decided on which function should be used to model the data, the next step is to find the values of the parameters of that function that describe the data best. To this problem we turn in the next section.

4.2.4 Methods for Fitting Psychometric Functions

There are different methods to find the best fitting curve to data. The methods differ with respect to the criterion by which "best-fitting" is defined. Here, we will discuss the most commonly used method for fitting the PF. It uses a maximum likelihood (ML) criterion, which defines the best-fitting PF to be that PF which would be most likely to replicate the experiment exactly as it was completed by the human observer. For a detailed discussion of the theory behind this procedure as well as a second, related procedure (Bayesian estimation) the reader is referred to Section B of this chapter. Here we will discuss, in the most general of terms, the basic idea behind the "maximum likelihood" method, and demonstrate how to perform a fit.

The example we will discuss uses the maximum likelihood criterion applied to fitting the Logistic function to a performance-based 2AFC task. First, we have to set up a series of vectors that contain the data. There are three that are required. As above, **StimLevels** provides the data values for the x axis. **NumPos** gives the number of trials in which the observer gave a correct response. **OutOfNum** gives the number of trials for each stimulus level. You can use any other name for any or all of these vectors if you like.

```
>>StimLevels = [0.01 0.03 0.05 0.07 0.09 0.11];
>>NumPos = [45 55 72 85 91 100];
>>OutOfNum = [100 100 100 100 100 100];
```

Next we have to specify the type of function we wish to use. The following command assigns the Logistic function to the variable PF as a MATLAB *inline* function. Other functions can be substituted for **Logistic**:

```
>>PF = @PAL_Logistic;
```

The following three commands set up parameters for the fitting procedure. In **paramsValues**, we give our initial guesses for α (threshold at 0.75 correct), and β (slope), as well as γ (guess rate), and λ (lapse rate). From a cursory inspection of the data we can see that performance is somewhere near 0.75 proportion correct when the stimulus level is about 0.05, so we enter 0.05 as the initial value of the first parameter, α. The slope parameter β is more difficult to estimate by inspection, but one can see that there is a large change in the number correct over a small change in stimulus level, so we put in a high number here, say 50. The guess rate γ is 0.5, and we will assume a lapse rate λ of 0 for this example. **paramsFree** specifies which of the four parameters α, β, γ, and λ are free parameters, that is parameters that the algorithm will attempt to find the best-fitting values for. We put 1 for a free parameter and 0 for a fixed parameter. Hence we have:

```
>>paramsValues = [0.05 50 .5 0];
>>paramsFree = [1 1 0 0];
```

Now we can run the curve fitting procedure as follows:

```
>>[paramsValues LL exitflag] = PAL_PFML_Fit(StimLevels, ...
NumPos,OutOfNum,paramsValues,paramsFree,PF)
```

The output is:

```
paramsValues =
0.0584 66.4520 0.5000 0

LL =
-273.4364

exitflag =
1
```

Note the new values of α and β in the vector **paramsValues**– these are the fitted values. The meaning of **LL** will be given in Section B of this chapter. The value of 1 for **exitflag** means that the fit was successful.

The function **PAL_PFML_Fit** finds the best-fitting parameters by way of an iterative search through different possible values of the parameters. It is possible to specify some of the characteristics of the search, for example the desired precision with which the parameter values are estimated. In Section B of this chapter we explain how to use this option (Section 4.3.3.1.2). In case the search characteristics are not specified by the user default values will be used, which for most practical purposes will be just fine. In Section B of this chapter it is also explained how you can limit the range of possible values that the lapse rate can assume such as to avoid impossible (e.g., negative) or improbable values.

Although it is the values of α and β that are important, it's nice to see what the fitted function looks like. The following creates a graph showing the data and the smooth fitted function.

```
>>PropCorrectData = NumPos./OutOfNum;
>>StimLevelsFine = [min(StimLevels):(max(StimLevels)- ...
min(StimLevels))./1000:max(StimLevels)];
>>Fit = PF(paramsValues,StimLevelsFine);
>>plot(StimLevels,PropCorrectData,'k.','markersize',40);
>>set(gca, 'fontsize',12);
>>axis([0 .12 .4 1]);
>>hold on;
>>plot(StimLevelsFine,Fit,'g-','linewidth',4);
```

The graph should look like Figure 4.6. Note that the graph plots proportion correct, not number correct against stimulus level. Note from the figure that the estimate of α ($\hat{\alpha} = 0.0584$) corresponds to the stimulus level at which the fitted function is at 0.75 proportion correct.

4.2.5 Estimating the Errors

Because the estimates of parameters α and β are based on a limited number of trials, they are indeed only estimates of the "true" values of α and β, the exact values

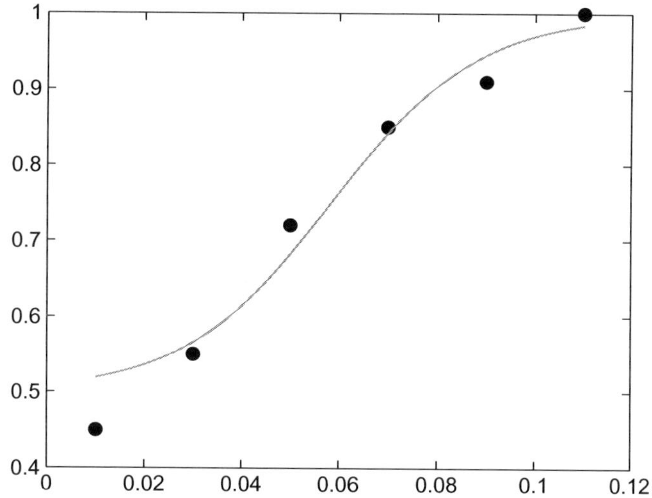

FIGURE 4.6 Plot generated by the code in the text.

of which we will never know. So even if we repeated the experiment under identical conditions, the estimated values of α and β would not come out exactly the same, due to the fact that we have a noisy brain. It would be useful, therefore, to obtain some sort of estimate of how much we might expect our estimates of α and β to vary from their true values. We could of course get a good idea of this by repeating our experiment, say 1,000 times, obtain estimates of α and β for each experiment, and then calculate the variance or standard deviation of the values across all experiments. Unfortunately we don't have the time to do this, but fortunately we can get a rough estimate of the likely variability in these parameters from just one set of data.

The preferred method for doing this is called "bootstrap analysis," and the details of the method are given in Section B of this chapter. The basic idea behind bootstrap analysis is that the computer randomly generates many sets of hypothetical data based on the actual experimental data obtained. Each new hypothetical data set is then fitted with the chosen function and estimates of α and β are obtained. The standard deviations of the α and β estimates across all the sets is then calculated, and these are the estimates of the errors on the parameters.

The function in the Palamedes toolbox that implements bootstrapping is **PAL_PFML_BootstrapParametric**. It requires that the PF fitting routine has already been run, and requires the same vectors as arguments as does the curve-fitting routine described in the previous section. Make sure that in **paramsValues** the parameter estimates as determined by **PAL_PFML_Fit** are used. One more argument is required. The argument **B** specifies how many simulated data sets are generated. The larger is **B**, the better the error estimate will be, but also the longer the routine will take to complete. Setting **B** to 400 should give an acceptable degree of accuracy on the error estimate, and it should also lead to an acceptable completion time. Here is an example implementation:

```
>>B = 400;
>>[SD paramsSim LLSim converged] =...
PAL_PFML_BootstrapParametric(StimLevels, OutOfNum, ...
paramsValues, paramsFree, B, PF);
```

In this example, the semicolon has been appended to the last line to prevent it from displaying the 400 estimates of α and β. To inspect the standard deviation of the estimates, type:

```
>>SD
```

An example output might be:

```
SD =
0.0035   11.7045   0   0
```

The four values are the estimates of the errors of our estimates of α, β, γ, and λ. Of course the values are only non-zero for the free parameters α and β. If you run the routine again and type out **SD**, the error estimates will be slightly different, because they will be based on a new set of simulated datasets. The larger the value of **B**, the closer will be the error estimate to the "true" error.

As its name suggests, the function **PAL_PFML_BootstrapParametric** performs what is known as a parametric bootstrap. An alternative is to perform a non-parametric bootstrap using the functions **PAL_PFML_BootstrapNonParametric**. An explanation of the distinction will have to wait for Section B of this chapter. Both **PAL_PFML_BootstrapParametric** and **PAL_PFML_BootstrapNonParametric** have a few optional arguments which will also be explained in Section B of this chapter.

4.2.6 Estimating the Goodness-of-Fit

The goodness-of-fit is a measure of how well the fitted PF accounts for the data. In general, if the data fall precisely along the fitted PF then this would be indicative of a good fit, whereas if the data points fall some way away from the fitted PF, this would indicate a bad fit. Goodness-of-fit measures of PFs can be useful for telling whether one type of fitting function is more appropriate than another. For example, a goodness-of-fit measure may guide one in deciding which of the different functions (Weibull, Logistic, etc.) is the better to use to model experimental data. A bad fit of a function may also be indicative of a high proportion of lapse trials when these are not accommodated for by the PF that is fitted.

The goodness-of-fit is determined by comparing two models statistically. For that reason, we will discuss the details of the procedure and the underlying rationale in Chapter 8, which deals with statistical model comparisons. For now, we will demonstrate the function in the Palamedes toolbox that performs a goodness-of-fit test and what to look for when deciding whether the PF fits your data well. The goodness-of-fit function in the Palamedes toolbox delivers two numbers: **Dev**, which stands for deviance, and **pDev**. The meaning of **Dev** will be explained in Chapter 8. The actual goodness-of-fit measure is given by **pDev**. **pDev** will always have a value between 0 and 1; the larger the value of **pDev**, the better the fit. By somewhat arbitrary convention, researchers agree that the fit is unacceptably poor if **pDev** is less than 0.05.

The goodness-of-fit routine in the Palamedes toolbox is **PAL_PFML_GoodnessOfFit**, and requires the best fitting parameter estimates found earlier by the PF fitting routine. The routine uses the same arguments as the error estimation routines described in the previous section, and these must of course all be defined. Here is an example implementation:

```
>>B = 1000;
>>[Dev pDev DevSim converged] = ...
PAL_PFML_GoodnessOfFit(StimLevels, NumPos, OutOfNum, ...
paramsValues, paramsFree, B, PF);
```

Note the semi-colon to prevent a full printout. Here also, **B** determines the number of simulations on which to base **pDev**. Once again, the higher the value assigned to **B**, the better the estimate of **pDev** will be, but the longer the routine will take to complete. After running the routine, type **Dev** and **pDev** to display the deviance and associated *p*-value:

```
Dev =
7.9773
pDev =
0.1013
```

pDev will have a slightly different value each time you run the function, because of the stochastic nature of the bootstrap.

4.2.7 Putting It All Together

Of course, one can put all the various components described in the previous sections together into a single MATLAB m-file. The Palamedes toolbox contains an m-file which does just that. The m-file is named **PAL_PFML_Demo**, and can be executed simply by typing its name at the command prompt. First, the program will prompt the user to select either a parametric or a non-parametric bootstrap. It then fits a Logistic function to some 2AFC data using the maximum likelihood criterion. The routine uses the same input vectors **StimLevels**, **NumPos**, **OutofNum**, **paramsValues**, **paramsFree** as in the above examples, and outputs the estimates of the six values described above: threshold α; slope β; SEs of both α and β; goodness-of-fit deviance; and *p*-value. Finally it generates a graph of the data and fitted function. In the m-file, all of the optional arguments to the functions are also demonstrated. These will be explained fully in Section B of this chapter. The routine simulates the experiment 4,400 times (for a total of 2,640,000 simulated trials!) so it will require a bit of time to complete. The modest laptop computer on which this sentence is being written just completed this routine in a little over a minute.

4.3 SECTION B: THEORY AND DETAILS

4.3.1 Psychometric Function Theories

As discussed in Section A, the psychometric function (PF) relates performance on a psychophysical task (e.g., probability of a correct response) to some characteristic of the stimulus (e.g., stimulus contrast). Following general consensus, we will denote performance on the task as a function of stimulus intensity x by $\psi(x)$. The shape of the PF is remarkably similar across a wide variety of tasks, and is typically well described by a sigmoidal function. More often than not, however, we are not directly interested in the measured performance in our experiment. Rather, we are interested in the sensitivity of the sensory mechanism underlying this performance. We will use $F(x; \alpha, \beta)$ (or simply $F(x)$) to symbolize the function describing the probability of correct stimulus detection or discrimination *by the underlying sensory mechanism* as a function of stimulus x. Section 4.3.2 discusses various models for $F(x)$ and its two parameters. $F(x)$ cannot be measured directly by psychophysical methods, and can only be inferred from performance as we measure it, $\psi(x)$.

Thus, it is worth considering, in some detail, how $\psi(x)$ and $F(x)$ might be related. We will consider this issue first in the context of "high-threshold theory," as this will lead us to the most commonly used expression of the relation between $\psi(x)$ and $F(x)$. We will then discuss how "signal detection theory" relates internal sensory mechanisms to the psychometric function. Other theories exist, and we refer the interested reader to Green and Swets (1966) for a more complete discussion.

4.3.1.1 High-threshold Theory

Let us imagine a simple two-interval forced-choice (2IFC) experiment in which the observer is presented on each trial with two intervals, one containing a stimulus, the other containing no stimulus. The stimulus interval is often denoted S (for signal), whereas the blank interval is denoted N (for noise). The observer is to determine, on each trial, which of the two intervals contained the stimulus.

Whether or not the sensory mechanism will detect the stimulus on any trial is determined by the amount of sensory evidence accumulated by the visual system as a result of the presentation of the stimulus. One may think of sensory evidence as some aggregate of the activity of a population of neurons selective for the to-be-detected stimulus. Due to external and internal noise, the amount of sensory evidence

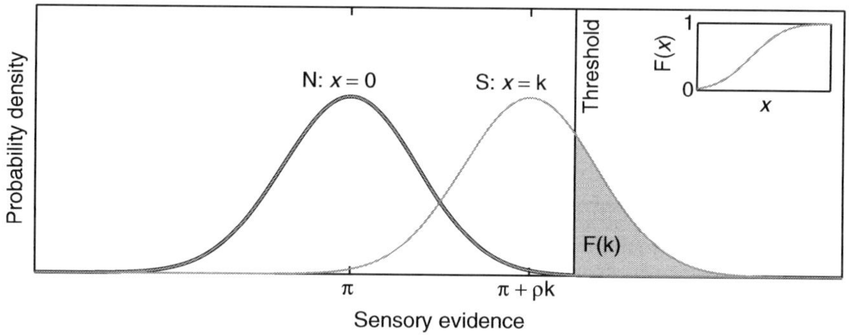

FIGURE 4.7 High-threshold theory and the psychometric function.

accumulated will fluctuate randomly from stimulus presentation to stimulus presentation, such that any given stimulus may give rise to varying amounts of sensory evidence. Let us assume that the mean amount of sensory evidence resulting from the presentation of a stimulus is a linear function of stimulus intensity x: $\mu(x) = \pi + \rho x$. Let us further assume that the random fluctuations in sensory evidence are distributed according to a normal distribution. The situation is depicted in Figure 4.7. This figure shows the probability density with which a stimulus at intensity k generates different amounts of sensory evidence. Also shown is the probability density associated with the interval that does not contain the stimulus (i.e., stimulus intensity $x = 0$). It can be seen in the figure that the probability density function of the stimulus with intensity $x = \mathrm{k}$ is centered around a higher average degree of sensory evidence. In general, increasing stimulus intensity will move the probability density function to higher degrees of expected sensory evidence.

According to high-threshold theory, the sensory mechanism will detect the stimulus when the amount of sensory evidence exceeds a fixed internal criterion or threshold. As its name implies, high-threshold theory assumes that the internal threshold is high. More specifically, the threshold is assumed to be high enough such that the probability that the threshold is exceeded when $x = 0$ (i.e., by noise alone) is effectively zero. This idea is reflected in the figure by the threshold being beyond the grasp of the $x = 0$ stimulus, thus the noise interval will never result in sensory evidence in excess of the threshold. It is this critical assumption of the high-threshold model which sets it apart from low-threshold theories. Another critical assumption of the high-threshold theory is that the decision process has no access to the exact amount of sensory evidence accumulated in case the threshold is not exceeded. The decision is based on binary information only: either the sensory evidence was in excess of the threshold, or the sensory evidence was not in excess of the threshold. This second critical assumption sets high-threshold theory apart from signal detection theory (Section 4.3.1.2 and Chapter 6). Given the assumptions we have made, function F(x), which describes the probability that the threshold will be exceeded by a stimulus of intensity x, will be the cumulative normal distribution. Function F(x) is shown in the inset in Figure 4.7.

The decision process is straightforward. Since the threshold cannot be exceeded by the noise interval, the sensory mechanism does not generate "false alarms." When the threshold is exceeded in one of the two intervals, it *must* have been because the signal was presented during that interval. In this situation the observer will identify the interval in which the stimulus was presented correctly. On those trials where the signal fails to exceed the threshold, however, the observer is left to guess which interval contained the signal. In this example, the observer will generate a correct response with a probability of 0.5 when the sensory evidence fails to exceed the internal threshold. In general, the probability of producing a correct response based on guessing is $1/m$ in an M-AFC task. As mentioned in Section A, the guess rate is conventionally denoted γ.

We need to make one more consideration before we are ready to consider how $\psi(x)$ and F(x) relate. In Section A, we mentioned that on each trial there is a small probability of an incorrect response which is independent of x. This probability is commonly referred to as the lapse rate, and is typically symbolized by λ. Lapses may occur because, for example, the observer did not witness the stimulus presentation (sneezes are often blamed). Another reason for a lapse might be a response error, in which the sensory mechanism may have identified the correct stimulus interval but, for some reason or another, the observer presses the incorrect response button. Anybody who has ever participated in psychophysical experiments will recognize that sometimes our thumbs seem to have a mind of their own.

We will illustrate how $\psi(x)$ and F(x) are related in Figure 4.8. This figure depicts the various series of events that lead to correct and incorrect responses. Starting at the top node, we separate the trials on which a lapse occurs (with probability λ^*) from those

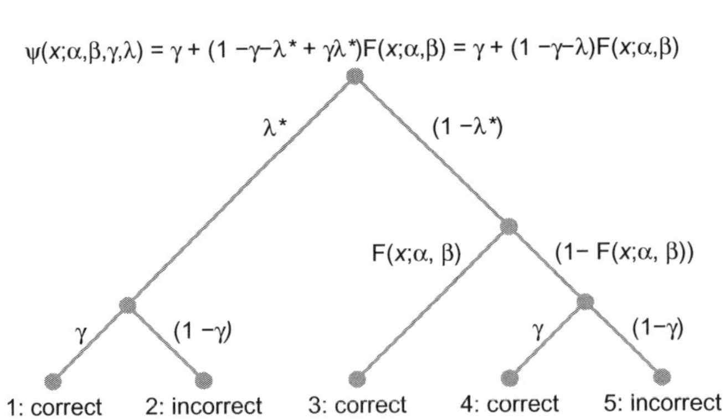

FIGURE 4.8 Relation between F(x; α, β) and $\psi(x; \alpha, \beta, \gamma, \lambda)$ according to high-threshold theory.

where a lapse does not occur ($1 - \lambda^*$). We use the symbol λ^* to distinguish this probability from the lapse rate, as defined above. Above, we defined the lapse rate as the probability of an *incorrect* response which is independent from x (which is the most common definition in the literature). However, in Figure 4.8 we use λ^* to symbolize the probability that the observer responds independently of stimulus intensity x (for example, resorts to a guess when the stimulus was not witnessed due to a sneeze). In such a situation, the response might still be correct with a probability equal to the guess rate, γ. This sequence actually corresponds to path 1 in the figure; the observer lapses (λ^*), then guesses correctly (γ). Since these two consecutive events are independent, the probability of this sequence is simply the product of the probabilities of the individual events (i.e., $\lambda^*\gamma$). In path 2, the observer lapses, resorts again to a guess but this time guesses wrong. The probability of this sequence is $\lambda^*(1 - \gamma)$.

On those trials where the observer does not lapse (with probability $1 - \lambda^*$), the sensory threshold will be exceeded with probability $F(x; \alpha, \beta)$. If this happens, the observer responds correctly and this completes path 3. The probability of this sequence of events equals $(1 - \lambda^*)F(x; \alpha, \beta)$. In path 4, the observer does not lapse ($1 - \lambda^*$), the sensory threshold is not exceeded $[1 - F(x; \alpha, \beta)]$, the observer resorts to a guess and guesses correctly (γ). The probability of this series of events is $(1 - \lambda^*)(1 - F(x; \alpha, \beta))(\gamma)$. Path 5 is identical to path 4, except that the observer guesses incorrectly. The probability with which this sequence occurs equals $(1 - \lambda^*)(1 - F(x; \alpha, \beta))(1 - \gamma)$.

Paths 1, 3, and 4 all result in a correct response. Since the five paths are mutually exclusive (only one can occur on any given trial), the probability of either one of these three paths occurring is the sum of the probabilities of the individual paths. Thus:

$$\psi(x;\alpha,\beta,\gamma,\lambda^*) = \lambda^*\gamma + (1 - \lambda^*)F(x;\alpha,\beta) + (1 - \lambda^*)(1 - F(x;\alpha,\beta))(\gamma),$$

which simplifies to:

$$\psi(x;\alpha,\beta,\gamma,\lambda^*) = \gamma + (1 - \gamma - \lambda^* + \lambda^*\gamma)F(x;\alpha,\beta) \qquad (4.2a)$$

Remember that the symbol λ^* refers to the probability with which the observer responds *independently* of the value of x, for example because the trial was not witnessed. On such trials a correct response may still occur by lucky guessing. More commonly, the following expression is used:

$$\psi(x;\alpha,\beta,\gamma,\lambda) = \gamma + (1 - \gamma - \lambda)F(x;\alpha,\beta) \qquad (4.2b)$$

The symbol λ used in this expression corresponds to the probability which is independent of x with which the observer will generate an incorrect response. The value $(1 - \lambda)$ corresponds to the upper asymptote of $\psi(x)$.

The parameter λ^* is more easily interpreted behaviorally than λ. For example, the value of λ^* for an observer who sneezes on every tenth trial and resorts to a guess

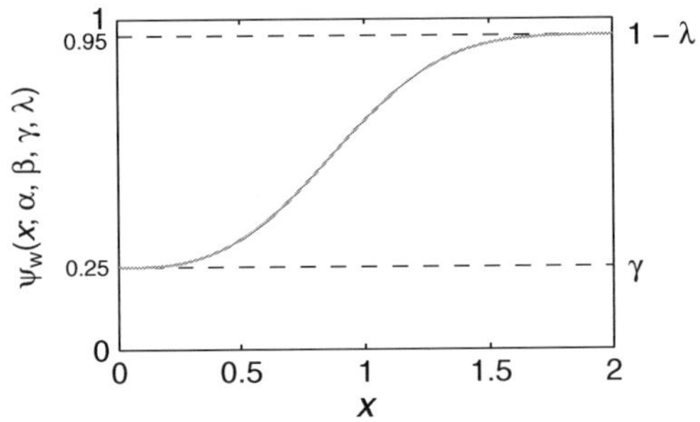

FIGURE 4.9 $\psi_W(x; \alpha, \beta, \gamma, \lambda)$, where F is modeled by the Weibull function $F_W(x; \alpha, \beta)$, threshold $\alpha = 1$, slope $\beta = 3$, guess rate $\gamma = 0.25$, and lapse rate $\lambda = 0.05$.

on those trials, is simply 1/10 (0.1). The value of λ, on the other hand, will also depend on the guess rate. In a 2AFC task, for example, the observer who sneezes on every tenth trial is expected to guess correctly on one half (γ) of the sneeze trials, and thus $\lambda = \lambda^*(1 - \gamma) = (0.1)(0.5) = 0.05$. Figure 4.9 displays $\psi_W(x; \alpha, \beta, \gamma, \lambda)$, where the subscript W indicates that function F is the Weibull function (Section 4.3.2), $\alpha = 1$, $\beta = 3$, $\gamma = 0.25$, and $\lambda = 0.05$.

4.3.1.2 *Signal Detection Theory*

The assumption, critical to high-threshold theory, that the amount of sensory evidence accumulated is unavailable to the decision process unless it exceeds some internal threshold stands in direct contrast to a central tenet of signal detection theory (SDT). According to SDT, there is no such thing as a fixed internal threshold. Instead, SDT makes the assumption that for all stimulus intensities x (including $x = 0$), the sensory mechanism generates a *graded* signal corresponding to the degree of sensory evidence accumulated. A decision process follows which considers the magnitude of this sensory evidence on both S and N intervals. If we again consider the two-interval forced-choice (2IFC) task from above, under the SDT framework both the noise and the signal intervals result in a degree of sensory evidence. This degree of evidence is again subject to external and internal noise, such that it will vary randomly from occasion to occasion even when identical stimuli are used. The situation is depicted in Figure 4.10. Under the SDT framework, the decision process has access to the degree of sensory evidence accumulated in both of the intervals. We may think of any presentation of a stimulus as a sample from the probability density function associated with the stimulus. Even in the absence of a stimulus, differing degrees of sensory evidence result, and we may think of the presentation of the noise interval as a sample from the probability density function associated with the noise stimulus. Thus, each of the two intervals on any trial gives rise to sensory evidence, the amount of which is known to the decision process.

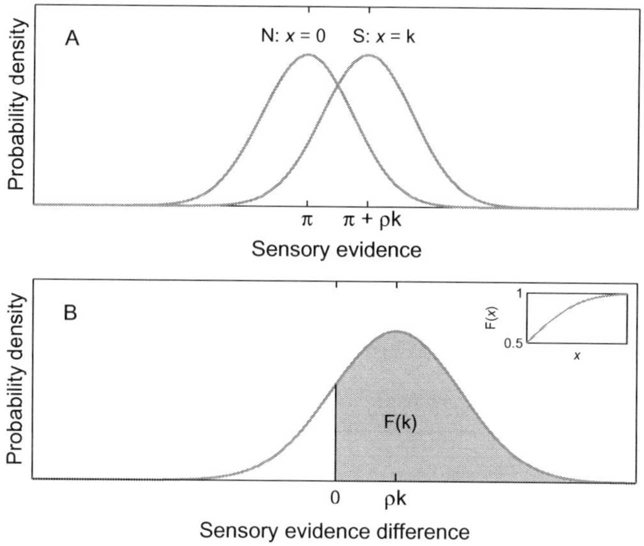

FIGURE 4.10 The relationship between SDT and the PF.

The decision rule of the observer is based directly on the relative amplitude of the two samples and is rather simple; the observer will report that the stimulus was presented in the interval from which the greater of the two samples was obtained. It is now easy to see how an incorrect response might arise. If we refer again to Figure 4.10, there is considerable overlap between the two functions. As a consequence, it is possible that the sensory activity sampled during the noise interval is greater compared to the activity sampled during the signal interval.

How is the probability of a correct response related to stimulus intensity? In order to generate a specific form of the PF we need to make a few assumptions, and we will do so now. Let us again assume that the mean sensory activity (μ) is a linear function of stimulus intensity level x: $\mu(x) = \pi + \rho x$, and that the variance is independent of stimulus level and equal to σ^2. In other words, the probability density function describing the sensory activity in noise intervals is normal, with mean π and variance σ^2: $N(\pi,\sigma^2)$ and that in stimulus intervals is $N(\pi + \rho x, \sigma^2)$. This situation is depicted schematically in Figure 4.10a.

Thus, each trial may be thought of as taking a sample from $N(\pi, \sigma^2)$ in the noise interval and a sample from $N(\pi + \rho x, \sigma^2)$ in the stimulus interval. We assume the observer utilizes a simple (but, given the assumptions, optimal) decision rule; the sample with the greater value was obtained in the stimulus interval. Thus, the response will be correct if the sample taken during the stimulus interval has a value greater than the sample taken during the noise interval. It will be convenient to rephrase this assumption as: if the difference between the sample value derived from the signal interval and the sample value derived from the noise interval exceeds zero, the response will be correct. It is well-known that the probability density of the difference between two normally distributed variables is itself normally distributed with mean equal to the difference in means of the individual distributions and variance of the difference equal to the sum of the variances of the individual distributions. Thus, specific to this example, the difference in sensory evidence will be distributed as $N(\rho x, 2\sigma^2)$. Figure 4.10b shows the density function for the difference in sensory evidence on a trial in which the signal is presented at intensity $x = k$. As noted, the stimulus interval will be correctly identified when the sampled difference in sensory activity exceeds zero. The probability with which this will occur corresponds to the shaded area in the figure. When the stimulus intensity equals zero the difference distribution will be $N(0, 2\sigma^2)$ and the probability that the difference score will exceed zero is 0.5. This makes sense, of course, at stimulus intensity 0, N and S are identical and the probability that S will produce a greater sensory activity than N is obviously 0.5. Increasing the stimulus intensity will move the difference distribution towards higher values, which corresponds to an increase in the probability that the difference score will exceed zero. Note that, under the assumptions of signal detection theory, the observer never truly guesses. The observer's response is on all trials determined by the relative degree of sensory evidence resulting in each of the stimulus intervals.

Under the assumptions made here, the PF will be the upper half of the cumulative normal density function (shown in the figure's inset). This shape of the PF is not encountered often. However, when we change our assumptions, especially with regard to the transducer function (which we above somewhat naïvely assumed to be linear) this would change the shape of the PF. The PF plotted in the figure's inset would take on the more commonly observed sigmoidal shape when we plot stimulus intensity x on a log scale, which is how stimulus intensities typically are plotted.

While the critical assumptions of high threshold theory have largely been discredited (e.g., Nachmias, 1981) in favor of those of SDT (for a detailed discussion of the issue see, for example, Swets (1961)), high-threshold theory lives on in the form of Equations 4.2a and 4.2b which are the most common formulaic expressions of the psychometric function used in the literature. Our nomenclature for two of the parameters of a PF is also based on high-threshold theory. Of course, the name "threshold" for the location parameter of a PF is based on high-threshold theory. Note that within the framework of the high-threshold theory the threshold defined as the amount of sensory evidence which needs to be exceeded before detection takes place is closely tied to the threshold as defined by the location parameter of a PF. Under the SDT framework there exists no fixed amount of sensory evidence beyond which the stimulus is detected and below which it is not. Nevertheless, we still refer to the location parameter as "threshold."

The name "guess rate" which we use for the lower asymptote (parameter γ) also has its basis in threshold theory. The assumption in high-threshold theory is that the stimulus is either detected or not, and if not, the observer guesses. The probability of a correct response on a trial in which the observer guesses corresponds to the lower asymptote. Quite naturally, under high-threshold theory the lower asymptote came to be referred to as the "guess rate." Today, we still refer to the lower asymptote as the guess rate. Moreover, in this text, we sometimes take even greater liberties. For example, where we should really say: "The amount of sensory evidence accumulated while sampling from the signal presentation happened to exceed the amount of sensory evidence accumulated while sampling from the noise presentation" we might say instead: "The observer guessed correctly."

4.3.2 Details of Function Types

Above we derived the generic formulation of the psychometric function (Section 4.3.1.1):

$$\psi(x;\alpha,\beta,\gamma,\lambda) = \gamma + (1 - \gamma - \lambda)F(x;\alpha,\beta) \tag{4.2b}$$

As discussed there, under the high-threshold detection model implied by this formulation, $F(x; \alpha, \beta)$ describes the probability of detection of the stimulus by the underlying sensory mechanism as a function of stimulus intensity x, γ corresponds to the guess rate (the probability of a correct response when the stimulus is not detected by the underlying sensory mechanism), and λ corresponds to the lapse rate (the probability of an incorrect response which is independent of stimulus intensity).

Several functions are in use for $F(x; \alpha, \beta)$. We list here the most commonly used functions. Examples of all these are shown in Figure 4.3. We will consistently use the symbol α to denote the location parameter (threshold), and the symbol β to denote the rate-of-change or slope parameter, even where this flies in the face of convention. We will also use expressions of F in which increasing values of β correspond to increasing slopes of F, even if this defies convention.

4.3.2.1 *Cumulative Normal Distribution*

The Cumulative Normal distribution is perhaps the most justifiable form of $F(x; \alpha, \beta)$ theoretically. If one assumes that the noise which underlies the variability of sensory evidence is a linear combination of many independent noise sources, the total resulting noise would be approximately normally distributed, by the well-known Central Limit Theorem (e.g., Hays, 1994). The Cumulative Normal distribution is given as:

$$F_N(x;\alpha,\beta) = \frac{\beta}{\sqrt{2\pi}} \int_{-\infty}^{x} \exp\left(-\frac{\beta^2(x-\alpha)^2}{2}\right) \tag{4.3}$$

with $x \in (-\infty, +\infty)$, $\alpha \in (-\infty, +\infty)$, $\beta \in (0, +\infty)$. No analytical solution to the integral is known, but the distribution may in practice be approximated by a numerical method. Parameter α corresponds to the threshold: $F_N(x = \alpha;\ \alpha,\ \beta) = 0.5$. Varying α while keeping β constant corresponds to a translation of the function. Parameter β corresponds to the reciprocal of the standard deviation of the normal distribution, and determines the slope of the PF. Since $F_N\ (x = 0;\ \alpha,\ \beta) > 0$ and $\lim_{x \to -\infty} F_N(x;\alpha,\beta) = 0$ for all values in the domains of α and β, the Cumulative Normal would be inappropriate in a task in which $x = 0$ corresponds to an absence of signal, unless x is log-transformed.

The function **PAL_CumulativeNormal** returns values of the psychometric function $\psi(x;\ \alpha,\ \beta,\ \gamma,\ \lambda)$ where $F(x;\ \alpha,\ \beta)$ is the cumulative normal distribution. Its usage is as follows:

```
>>pcorrect = PAL_CumulativeNormal([alpha beta gamma ...
lambda], StimLevels)
```

where **alpha**, **beta**, **gamma** and **lambda** correspond to the parameter values characterizing the PF, and **StimLevels** is a scalar, vector, or matrix containing values at which the PF is to be evaluated. The user may opt to provide values only for alpha and beta, in which case gamma and lambda are assumed to be zero. In that case, the function simply returns the values of F_N (that is: $\psi_N(x;\ \alpha,\ \beta,\ \gamma = 0,\ \lambda = 0) = F_N(x;\ \alpha,\ \beta)$). The return argument is a matrix of same size as **StimLevels**, which contains the function evaluations at each of the values contained in **StimLevels**. For example, to produce a plot of a PF in which $F(x;\ \alpha,\ \beta)$ is the cumulative normal with threshold $\alpha = 2$ and slope $\beta = 1$, the guess rate $\gamma = 0.5$, and the lapse rate $\lambda = 0$, type:

```
>>StimLevels = [0:.01:4];
>>pcorrect = PAL_CumulativeNormal([2 1 .5 0], StimLevels);
>>plot(StimLevels, pcorrect);
```

4.3.2.2 Logistic

The logistic function is given as:

$$F_L(x;\alpha,\beta) = \frac{1}{1 + \exp(-\beta(x - \alpha))} \tag{4.4}$$

with $x \in (-\infty,\ +\infty)$, $\alpha \in (-\infty,\ +\infty)$, $\beta \in (0,\ +\infty)$. Parameter α corresponds to the threshold: $F_L\ (x = \alpha;\ \alpha,\ \beta) = 0.5$, parameter β determines the slope of the PF. The logistic function is a close approximation to the Cumulative Normal distribution (after a linear transformation of the slope parameter β: $\beta_L = 1.7 \times \beta_N$). An advantage of the Logistic function over the Cumulative Normal is that the former can be evaluated analytically while the latter can not. For the same reasons as outlined for the Cumulative Normal distribution, the Logistic is inappropriate when a stimulus intensity $x = 0$ corresponds to an absence of signal, unless x is log-transformed.

Use of the toolbox function **PAL_Logistic** is analogous to that of **PAL_CumulativeNormal**.

4.3.2.3 *Weibull*

The Weibull function is given as:

$$F_W(x;\alpha,\beta) = 1 - \exp\left(-\left(\frac{x}{\alpha}\right)^{\beta}\right) \tag{4.5}$$

with $x \in [0, +\infty)$, $\alpha \in (0, +\infty)$, $\beta \in (0, +\infty)$. Threshold α corresponds to $F_W(x = \alpha; \alpha, \beta) = 1 - \exp(-1) \approx 0.6321$, parameter β determines the slope in conjunction with α. That is, changing the value of α will alter the slope of the function, even when β is held constant. However, when plotted against log x, a change in α will result in a translation of the function when β is held constant. $F_W(0; \alpha, \beta) = 0$ for all α, β. Note that since the domain of x includes positive numbers only, the Weibull function should not be used when x is measured in logarithmic units; the Gumbel function should be used instead.

Quick (1974) has argued that the Weibull function provides an excellent approximation to the PF when performance is determined by probability summation among channels with normally distributed noise. Use of the toolbox function **PAL_Weibull** is analogous to that of **PAL_CumulativeNormal**.

4.3.2.4 *Gumbel (Also Known as Log-Weibull)*

$$F_G(x;\alpha,\beta) = 1 - \exp(-10^{\beta(x-\alpha)}) \tag{4.6}$$

with $x \in (-\infty, +\infty)$, $\alpha \in (-\infty, +\infty)$, $\beta \in (0, +\infty)$. Threshold α corresponds to $F_G(x = \alpha; \alpha, \beta) = 1 - \exp(-1) \approx 0.6321$. The Gumbel function is the analog of the Weibull function when a log-transform on x is used. For that reason, in the literature the Gumbel function is often referred to as the log-Weibull function or, somewhat confusingly, simply as the Weibull function.

Use of the toolbox function **PAL_Gumbel** is analogous to that of **PAL_CumulativeNormal**.

4.3.2.5 *Hyperbolic Secant*

$$F_{HS}(x;\alpha,\beta) = \frac{2}{\pi}\tan^{-1}\exp\left(\frac{\pi}{2}\beta(x-\alpha)\right) \tag{4.7}$$

with $x \in (-\infty, +\infty)$, $\alpha \in (-\infty, +\infty)$, $\beta \in (0, +\infty)$. Threshold α corresponds to $F_{HS}(x = \alpha; \alpha, \beta) = 0.5$. Use of the Hyperbolic Secant is relatively rare in the psychophysical literature. We include it here for completeness.

Use of the toolbox function **PAL_HyperbolicSecant** is analogous to that of **PAL_CumulativeNormal**.

4.3.2.6 *Inverse Psychometric Functions*

The inverse PFs are also included in the Palamedes toolbox. In general, given a function f(x), its inverse function $f^{-1}(x)$ is defined such that $f^{-1}[f(x)] = x$. Thus, whereas a PF gives the probability of a correct response as a function of a specific stimulus level, an inverse PF gives the stimulus level which corresponds to a specific probability of a correct response. The general use of these functions in Palamedes is as follows:

```
>>StimLevels = PAL_inverse[NameOfFunction](params, pcorrect);
```

For example, to find the stimulus levels at which a logistic function with $\alpha = 1$, $\beta = 2$, $\gamma = 0.5$, and $\lambda = 0.01$ has the values 0.65, 0.75, and 0.85, type:

```
>>StimLevels = PAL_inverseLogistic([1 2 .5 0.01], [0.65 ...
0.75 0.85]);
>>StimLevels
```

MATLAB reports:

```
StimLevels =
0.5908   1.0204   1.4581
```

4.3.2.7 *The Spread of Psychometric Functions*

We have used β in all of the forms of the PF above to symbolize the parameter which determines the steepness of the PF. Because β affects the steepness of the PF, it is often referred to as the slope parameter, or simply "the slope" of the PF. This is not entirely proper, as β does not directly correspond to the slope of the function as it is defined in calculus. Moreover, values of β cannot be compared directly between the different forms of PF. For example, a Cumulative Normal function with $\beta = 2$ is much steeper compared to a Logistic function with $\beta = 2$. A common measure related to the steepness of PFs is the "spread" (or "support"). The spread will actually have an inverse relation to the slope of the PF. Remember that all of the PFs display asymptotic behavior. That is, as stimulus intensity increases ψ asymptotes towards $1 - \lambda$, but will never actually attain that value. Similarly, as stimulus intensity decreases, ψ asymptotes towards γ (with the exception of the Weibull whose value at $x = 0$ actually equals γ). As such, we cannot define spread as the range of stimulus intensities within which ψ goes all the way from the lower asymptote γ to the upper asymptote $1 - \lambda$. Instead, we pick an arbitrary number δ (e.g., 0.01) and define the spread to be that stimulus range within which ψ goes from $\gamma + \delta$ to $1 - \lambda - \delta$. Formally, if we let σ symbolize spread:

$$\sigma = \psi^{-1}(1 - \lambda - \delta;\alpha,\beta,\gamma,\lambda) - \psi^{-1}(\gamma + \delta;\alpha,\beta,\gamma,\lambda) \tag{4.8}$$

where $\psi^{-1}(y; \alpha, \beta, \gamma, \lambda)$ is the inverse of the psychometric function $\psi(x; \alpha, \beta, \gamma, \lambda)$. The value of δ must, of course, be between 0 and $(1 - \gamma - \lambda)/2$. The Palamedes function **PAL_spreadPF** gives the spread of a PF. We demonstrate use of **PAL_spreadPF** by example. In order to find the spread (using $\delta = 0.01$) of a logistic function characterized by $\alpha = 2$, $\beta = 3$, $\gamma = 0.5$, and $\lambda = 0.01$ we type:

```
>>params = [2 3 0.5 0.01];
>>delta = 0.01;
>>spread = PAL_spreadPF(params, delta, 'logistic')
```

MATLAB returns:

```
spread =
2.5808
```

Instead of using **'logistic'** as an argument, we may use **'cumulativenormal'**, **'weibull'**, **'gumbel'**, or **'hyperbolicsecant'**. We do not need to worry about capitalization of this argument as case is ignored by **PAL_spreadPF**.

4.3.3 Methods for Fitting Psychometric Functions

The raw data resulting from a psychophysical experiment are the proportions of correct responses measured at a number of different stimulus intensities x. Each of these is based on a limited number of trials, and hence is only an estimate of the true probability with which the observer generates a correct response. We assume that the true probabilities of a correct response as a function of x are given by Equation 4.2b. Since, in most situations, we are interested in describing the properties of the underlying sensory mechanism we are interested only in determining the values of the threshold (α) and slope (β) of the function $F(x; \alpha, \beta)$. The guess rate (γ) is usually known ($1/m$ in an M-AFC task). The lapse rate (λ) is unknown, but is considered to be a nuisance parameter as it tells us nothing about the sensory mechanism *per se*. We may, of course, attempt to estimate it, but when we do so it is to improve our estimates of α and β (although situations might be imagined where the lapse rate is of interest for its own sake, in which case α and β might be considered nuisance parameters). We might also be interested in the precise shape of the function $F(x; \alpha, \beta)$. For example, it might be of theoretical interest to determine whether the Weibull function, say, provides a better fit to our data compared to the Logistic function.

We have a number of methods available to us to find the best-fitting psychometric function (PF) to our data. These methods differ ultimately with respect to the criterion by which "best-fitting" is defined. We will discuss two different methods in some detail. The first method uses the criterion of maximum likelihood to define best-fitting, the second uses a Bayesian criterion.

4.3.3.1 *Maximum Likelihood Criterion*

4.3.3.1.1 A Simple 1-Parameter Example

Let us start with a simple example to introduce the concept of "likelihood." Imagine that we have a coin and we wish to estimate the parameter corresponding to the probability that our coin lands "heads" on any given flip of the coin. We will designate this parameter α. We perform a (rather modest) experiment which consists of flipping the coin 10 times. After each flip, we note whether it landed "heads" (H) or "tails" (T). The results of our ten trials are respectively:

HHTHTTHHTH

The likelihood function associated with our parameter of interest is:

$$L(a \mid \mathbf{y}) = \prod_{k=1}^{N} p(y_k \mid a) \tag{4.9}$$

(e.g., Hoel, Port, & Stone, 1971), where a is a potential value for our parameter of interest, $p(y_k \mid a)$ is the probability of observing outcome y on trial k given or (perhaps more appropriately in this context) "assuming" value a for our parameter and N is our total number of trials (here, $N = 10$). In our example, it is obvious that $p(y_k = H \mid a) = a$ and $p(y_k = T \mid a) = 1 - a$. Equation 4.9 utilizes what is known as the multiplicative rule in probability theory (sometimes referred to as the "and rule"), which states that the probability of observing two or more events is equal to the product of the probabilities of the individual events when the events are independent. Thus, the likelihood associated with, say, $a = 0.4$ is:

$$\begin{aligned} L(0.4 \mid \mathbf{y}) &= \prod_{k=1}^{N} p(y_k \mid 0.4) \\ &= p(y_1 = H \mid 0.4) \cdot p(y_2 = H \mid 0.4) \cdot p(y_3 = T \mid 0.4) \cdot \ldots\ldots\ldots \cdot p(y_{10} = H \mid 0.4) \\ &= 0.4 \cdot 0.4 \cdot (1 - 0.4) \cdot \ldots \cdot 0.4 = 0.4^6 \cdot (1 - 0.4)^4 = (0.4)^6 \cdot (0.6)^4 \\ &\approx 0.000531 \end{aligned}$$

In words, the likelihood $L(0.4 \mid \mathbf{y})$ is calculated as the probability of observance of the outcome of 10 flips, *exactly as they occurred in our experiment*, from a coin for which α is known to be 0.4. Importantly, contrary to intuitive appeal perhaps, it would be inappropriate to consider $L(a \mid \mathbf{y})$ a probability, although we calculate it as such. In the context of our experiment we cannot think of $L(a \mid \mathbf{y})$ as the probability of obtaining our experimental outcome, simply because our experiment is over and there is no uncertainty (anymore) as to the outcome. Thus, our obtained value for $L(0.4 \mid \mathbf{y})$ does not give us information about the outcome of our completed experiment. Rather, we calculate it to gain information about the value for α.

Our obtained value for $L(0.4 \mid \mathbf{y})$, however, is also most certainly not the probability of parameter α having the value 0.4. Thus, $L(a \mid \mathbf{y})$ is not the probability of anything, and using the term "probability" would be inappropriate. Instead, we use the term "likelihood."

The likelihood function is a function of a and we may calculate $L(a \mid \mathbf{y})$ for any value of a. Figure 4.11 plots $L(a \mid \mathbf{y})$ as a function of a across the range $0 \leq a \leq 1$ (since a represents a probability, it must have a value within this range). As the term implies, the maximum likelihood estimate of parameter α is that value of a that maximizes the likelihood function $L(a \mid \mathbf{y})$. In our example, $L(a \mid \mathbf{y})$ is at maximum when a equals 0.6. Thus, $\hat{\alpha} = 0.6$ is our maximum likelihood estimate of α.

It may be noted that Equation 4.9 calculates the probability of observance of an exact "ordered" sequence of, for example, heads and tails (more generally, "successes" and "failures"), on a series of N independent trials. Some authors opt to include the binomial coefficient in Equation 4.9 to arrive at the probability of observing the numbers of successes and failures *in any order*:

$$L(a \mid \mathbf{y}) = \frac{N!}{m!(N-m)!} \prod_{k=1}^{N} p(y_k \mid a) \tag{4.10}$$

where m is the number of successes (heads). However, since the value of the binomial coefficient is determined entirely by the observed outcome of the experiment and does not depend on a, inclusion of the binomial coefficient amounts merely to a linear rescaling of the values of $L(a \mid \mathbf{y})$, and thus will not affect our estimate of α.

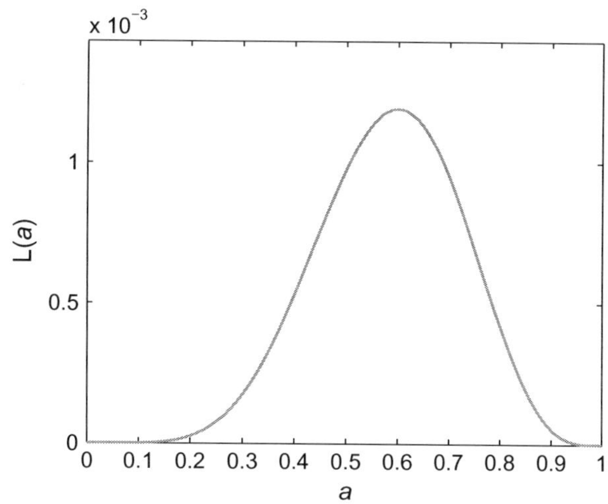

FIGURE 4.11 Plotted is the likelihood as a function of a, the (hypothesized) value for the probability of observance of heads on any given flip of our coin with unknown α.

One should not be discouraged by the small values of $L(a\,|\,\mathbf{y})$ obtained. Even the likelihood for our best estimate of α (0.6) amounts to a mere 0.0012. In other words, a coin for which $\alpha = 0.6$ would, when flipped ten times, have a probability of only 0.0012 of generating the sequence of outcomes we have observed in our experiment. This seems so unlikely that it might be tempting to conclude that our estimate $\hat{\alpha} = 0.6$ is not a very good one! This conclusion would be inappropriate, however. In effect, we have witnessed the occurrence of an event and have calculated *post hoc* the probability that this event *would* occur under certain assumptions (specifically, for a range of values of a). However, as argued above, this probability can be interpreted neither as the probability of our experiment resulting in the observed sequence, nor as the probability of α having the value of 0.6.

4.3.3.1.2 The Psychometric Function and the Likelihood Function

Typically, we wish to estimate two parameters of the psychometric function: its threshold (α); and its slope (β). Thus the likelihood function is now a function of two parameters and becomes:

$$L(a,b\,|\,\mathbf{y}) = \prod_{k=1}^{N} p(y_k\,|\,x_k;a,b), \tag{4.11}$$

where $p(y_k\,|\,x_k;\,a,\,b)$ is the probability of observance of response y (in an M-AFC task typically "correct" or "incorrect") on trial k given stimulus intensity x_k and assuming threshold $\alpha = a$ and slope $\beta = b$ of the psychometric function. Let us again imagine a modest experiment: an observer is to decide which of two temporal intervals contains a low-intensity visual stimulus. Five stimulus intensities x are used and the observer is tested four times at each of the five stimulus intensities. Table 4.2 presents the responses observed for each of the twenty trials (1: correct; 0: incorrect). Also listed for each response is the likelihood for two, somewhat arbitrary, assumed psychometric functions. These individual trial likelihoods are, of course, simply equal to $\psi(x_k;\,a,\,b,\,\gamma,\,\lambda)$ if the response is correct or $1 - \psi(x_k;\,a,\,b,\,\gamma,\,\lambda)$ if the response is incorrect. One of the functions is characterized by $a = 1$, $b = 1$, the other by $a = 10$, $b = 1$. We assume the guess rate γ equals 0.5, and the lapse rate λ equals 0. Since γ and λ are fixed we will denote the probability of a correct response as $\psi(x_k;\,a,\,b)$ for purposes of brevity. The two PFs for which we explicitly calculate the likelihoods are shown in Figure 4.12. Following Equation 4.11, the likelihood based on all twenty responses is simply calculated as the product of the likelihoods based on all individual trials. These overall likelihoods are also listed in the table. The interpretation of the likelihood here is analogous to that in the previous section. For example, the value $L(a = 1,\,b = 1\,\,|\,\mathbf{y}) = 4.078 \times 10^{-5}$ can be interpreted as the probability that an observer whose true underlying PF is characterized by $\alpha = 1$ and $\beta = 1$ would generate the exact sequence of responses as that produced

TABLE 4.2 The log-transformed stimulus level [log(x)], the observed outcome of the trial (y, 0 = incorrect, 1 = correct), the probability of a correct response for two assumed PFs [ψ (x_k; $a = 1, b = 1$) and ψ (x_k; a = 10, b = 1)], and the likelihood of each of the observed outcomes for both assumed PFs [$p(y_k \mid x_k; a = 1, b = 1)$ and $p(y_k \mid x_k; a = 10, b = 1)$] are shown for each of 20 trials (also shown are the likelihoods for the two PFs considered across the entire experiment [$L(1,1 \mid \mathbf{y})$ and $L(10,1 \mid \mathbf{y})$])

k	log(x)	y	$\psi(x_k; a = 1, b = 1)$	$p(y_k \mid x_k; a = 1, b = 1)$	$\psi(x_k; a = 10, b = 1)$	$p(y_k \mid x_k; a = 10, b = 1)$
1	−2	1	0.5596	0.5596	0.5237	0.5237
2	−2	0	0.5596	0.4404	0.5237	0.4763
3	−2	1	0.5596	0.5596	0.5237	0.5237
4	−2	0	0.5596	0.4404	0.5237	0.4763
5	−1	0	0.6345	0.3655	0.5596	0.4404
6	−1	1	0.6345	0.6345	0.5596	0.5596
7	−1	1	0.6345	0.6345	0.5596	0.5596
8	−1	1	0.6345	0.6345	0.5596	0.5596
9	0	1	0.7500	0.7500	0.6345	0.6345
10	0	1	0.7500	0.7500	0.6345	0.6345
11	0	0	0.7500	0.2500	0.6345	0.3655
12	0	1	0.7500	0.7500	0.6345	0.6345
13	1	1	0.8655	0.8655	0.7500	0.7500
14	1	1	0.8655	0.8655	0.7500	0.7500
15	1	1	0.8655	0.8655	0.7500	0.7500
16	1	0	0.8655	0.1345	0.7500	0.2500
17	2	1	0.9404	0.9404	0.8655	0.8655
18	2	1	0.9404	0.9404	0.8655	0.8655
19	2	1	0.9404	0.9404	0.8655	0.8655
20	2	1	0.9404	0.9404	0.8655	0.8655
				$L(1,1 \mid \mathbf{y}) = 4.078 \times 10^{-5}$		$L(10,1 \mid \mathbf{y}) = 2.654 \times 10^{-5}$

by our observer. Again, we should not be discouraged by the minute value of the likelihood: even if our observed proportion of correct for each of the stimulus levels would be perfectly predicted by the PF, our likelihood would be rather minute.

Of course, we may calculate the likelihood for any particular combination of a and b. Figure 4.13 presents a contour plot of $L(a, b \mid \mathbf{y})$ as a function of log a and log b

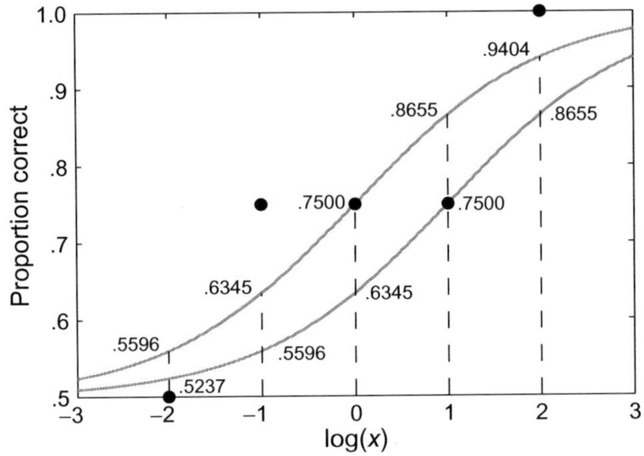

FIGURE 4.12 Shown is the probability correct (ψ) for two PFs (both are Logistic, one characterized by $\alpha = 1, \beta = 1$, the other by $\alpha = 10, \beta = 1$) as a function of log stimulus intensity [log(x)]. Also shown are the results of our experiment as the proportion of correct responses for each of the five stimulus intensities used.

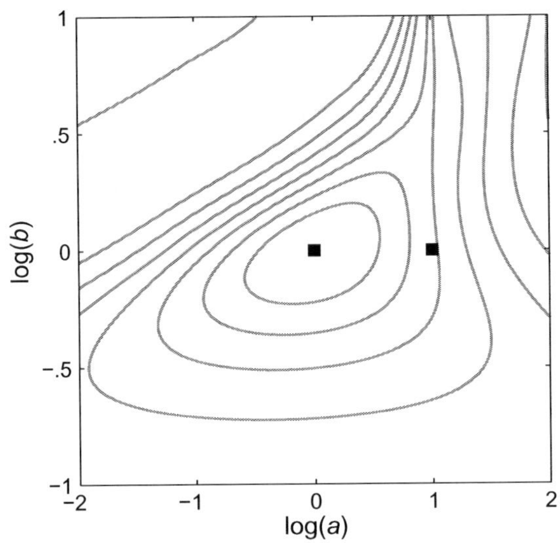

FIGURE 4.13 Shown is a contour plot of the likelihood function as a function of assumed threshold a and slope b. Square symbols correspond to the two PFs considered in Table 4.2 and shown in Figure 4.12. Contour lines correspond to $L(a, b \mid \mathbf{y}) = 0.5 \times 10^{-5}, 1 \times 10^{-5}, \ldots, 4 \times 10^{-5}$.

across the ranges $\log(a) \in [-2, 2]$ and $\log(b) \in [-1, 2]$. The two specific PFs whose likelihoods are calculated in Table 4.2 are indicated in the figure by the square symbols. Analogous to the previous one-parameter coin-flipping experiment described above, the maximum likelihood estimates of the threshold and slope of the PF are those values of a and b that maximize $L(a,b \mid \mathbf{y})$.

In practice, we perform our calculations using log-transformed probabilities:

$$LL(a,b \mid \mathbf{y}) = \sum_{k=1}^{N} \log_e p(y_k \mid x_k; a,b) \tag{4.12}$$

where $LL(a, b \mid \mathbf{y})$ is the "log likelihood" and $p(y_k \mid x_k; a, b)$ is as defined above. Since the log-transform is monotonic, the maximum value of $LL(a, b \mid \mathbf{y})$ and that of $L(a, b \mid \mathbf{y})$ will occur at corresponding values of a and b. One reason for performing our calculations on the log-transform is that, with increasing N, likelihoods become vanishingly small and may, in many practical applications, become too small to be represented (other than as "0"), as (64-bit) data type "double" in MATLAB (the smallest positive number that can be represented by a double is 2.22507×10^{-308}).

Note that, in the above, the probability of a correct response for the observer on any trial is assumed to be a function only of stimulus intensity. In other words, the probability of a correct response given stimulus intensity x is assumed to be identical, regardless of whether it is the first trial an observer performs, or the last trial, or any in between. We will call this the "assumption of stability." Due to practice and fatigue effects, the assumption of stability is almost certainly never strictly true. Another assumption which is implicitly made, when we assume that the probability of a correct response is a function of stimulus intensity only, is what we refer to as the "assumption of independence." The assumption of independence states that whether an observer responds correctly on any trial is affected by stimulus intensity only, and is not affected by whether the observer responded correctly or incorrectly on any other trial. In practice, the assumption of independence is also almost certainly never true. An example of a violation of this assumption would be if, after a series of incorrect responses, an observer becomes frustated and increases her concentration and attention in order to optimize the probability of getting the next trial correct.

We will not attempt here to visualize a three- or four-dimensional likelihood function, but we may include the guess rate and/or the lapse rate as parameters in the log likelihood function in order to estimate them as well. In an M-AFC task the guess rate is known, but situations can be imagined where the guess rate is not known and it may need to be estimated. In practice, we do not know what the lapse rate is. In theory, we can estimate the lapse rate, but we might run into problems when we attempt to do so. Some of these problems will be discussed below. In practice, many researchers implicitly assume a lapse rate equal to zero in their fitting procedure. When we do this, however, even a single lapse during an experiment may have a significant effect on our estimates of the threshold and slope parameters. We will elaborate a bit on the issues involved.

Imagine a 2AFC experiment in which we aim to estimate the threshold and slope of a psychometric function. The method of constant stimuli is used with stimulus values $x = -4, -3, \ldots, 4$. Let us assume that the observer's actual F(x) is the logistic function with $\alpha = 0$ and $\beta = 2$. We will consider here the effect of a lapse at a single trial in which the stimulus intensity equals 4. Under the assumption that our observer's lapse rate equals 0, the probability that this observer would generate an incorrect response at the highest stimulus intensity is extremely low: $\psi_L(x = 4; \alpha = 0, \beta = 2, \gamma = 0.5, \lambda = 0) = 0.99983$. Remember that the likelihood associated with a particular PF may be thought of as the probability that this PF would produce the exact sequence of responses (including this particular trial) that was in fact observed. Thus, when a lapse does occur on this trial and an incorrect response is produced, the likelihood associated with this PF will be severely deflated as a consequence. Specifically, due to the lapse, the likelihood will be multiplied by $1 - 0.99983$ or 0.00017 based on this one trial's observed response. However, had the observer not lapsed and responded correctly, the likelihood would be multiplied by 0.99983. Thus, the occurrence of the lapse affected the likelihood by a factor of $0.00017/0.99983 = 0.00017$ relative to the situation in which the lapse had not occurred.

Let us also examine the effect of this lapse on the likelihood associated with a different PF: that characterized by $\alpha = 1$ and $\beta = 1$. $\psi_L(x = 4;\ \alpha = 1,\ \beta = 1,\ \gamma = 0.5, \lambda = 0) = 0.9763$. In other words, this PF may, on occasion, produce an incorrect response when $x = 4$. Following the same logic as above, the lapse affected the likelihood of this PF by a factor of $0.0237/0.9763 = 0.0243$. Thus, the effect of the lapse is much less dramatic for this PF compared to the PF discussed above. In general, the effect of lapses on likelihoods will vary systematically across the parameter space. Generally speaking, lapses that occur at high stimulus intensities will severely suppress the likelihoods associated with PFs that are highly unlikely to produce incorrect responses at these stimulus intensities. Lapses affect the likelihoods associated with PFs that allow for occasional incorrect responses at these stimulus intensities to a much lesser degree. In general, then, lapses occurring at high stimulus intensities will lead to an overestimation of the threshold parameter and an underestimation of slope parameters.

How can we minimize the effect of lapses? First, we should try to avoid the occurrence of lapses as much possible. However, this is easier said than done. Anybody who has ever participated in psychophysical experiments will know that lapses are impossible to avoid entirely. There are a few procedural methods we could consider to avoid lapses due to the observer being trigger-happy. Due to the repetitive nature of psychophysical testing, observers naturally automate responding to some extent, and occasionally a response button may be pressed without conscious intent. In an *M*-IFC task, observers may have made up their mind and initiated a response after having witnessed only the first interval. Such issues may be alleviated simply by disallowing early responses. One may even consider sounding an unpleasant beep when the observer responds too early. Meese and Harris (2001) have proposed a procedure which allows observers to reconsider responses that are the result of being trigger-happy. In this procedure, responses are recorded only when a response button is released (rather than depressed).

Given that lapses will occur, we have several options to minimize their effect when they do occur. One option we have is to free the lapse parameter during our fitting procedure. As it turns out, however, since lapses are rare, the estimate of our lapse rate is typically not very accurate (e.g., Wichmann & Hill, 2001). From our own experience, the estimate for the lapse rate will often have a negative value. Also, in general, the more parameters we attempt to estimate the more likely it is that our fitting procedure will fail. For example, when we use a search algorithm to find the maximum likelihood it will be more likely that we find a local maximum rather than the absolute maximum in the likelihood function when we increase the number of free parameters to be included in the fit.

We may avoid negative values for the estimate of our lapse rate and also reduce the probability that fits will fail by constraining the lapse rate to have a value within a narrow range of reasonable values, say $0 \leq \lambda \leq 0.06$ (e.g., Wichmann & Hill, 2001). However, when we do so, estimates of lapse rates tend to end up at either limit of the range in many realistic situations (e.g., García-Pérez & Alcalá-Quintana, 2005). Whereas setting the lower limit at zero is quite natural, the upper limit is inherently somewhat arbitrary. Also, one might argue that when we recognize the need to constrain the lapse rate to a range of values, we implicitly acknowledge that our data are insufficient to result in a reliable estimate of the lapse rate.

If we do wish to estimate our lapse parameter, Treutwein (1995) suggests we include a relatively high proportion of trials in our experiment at high stimulus intensities. Responses at high stimulus intensities give much information as to the lapse rate. This strategy is, however, rather expensive in terms of the number of trials required. Since we can expect lapses to occur on only a few percent of trials, we would need many trials at high stimulus intensities to acquire an accurate estimate of the lapse rate. However, responses at these stimulus intensities give us little information as to the threshold and slope parameters.

In a typical psychophysical experiment, observers will test under several experimental conditions. For example, sensitivity to a particular stimulus might be determined both with and without adaptation. Since the probability with which the observer will lapse is, by definition, independent of the stimulus, we can assume that the true lapse rate is identical across all conditions of an experiment. This raises the possibility of estimating a single lapse rate based on the entire experiment. The lapse rate will be estimated much more accurately, since it will be based on many more trials compared to estimates of lapse rates based on the results in a single condition. In Chapter 8 we will show how we can estimate a single lapse rate across several conditions of an experiment while still being able to estimate individual thresholds and slopes for each of the conditions.

If we do not wish to estimate our lapse rate, we should avoid extremely high stimulus intensities. As shown above, when the probability of detection by the underlying mechanism of a stimulus is near 1, but the response is incorrect due to a lapse, the likelihood associated with the true PF is severely deflated. However, when the stimulus intensity is very low and the probability of detection by the underlying mechanism (i.e., $F(x)$) is near 0, it does not matter much whether the observer lapses or not. The observer performs around chance level either way. In general, the higher the stimulus intensity, the greater the effect of a lapse will be on our threshold and slope parameter estimates. Also, the cost associated with avoiding high stimulus intensities is very low, because one should avoid extremely high stimulus intensities anyway, since responses here provide little information as to the values of our threshold and slope parameters (Chapter 5 will discuss this issue in more detail).

A final strategy to minimize lapse effects is to fix the lapse rate during the fitting procedure at a small but non-zero value, such as 0.02. While lapses will still affect our parameter estimates, the occurrence of a lapse at a stimulus intensity where $F(x)$ is near 1 will not have the catastrophic effect it would have if we fixed the lapse rate at 0. Wichmann and Hill (2001) have systematically studied the effect of lapses on the estimates of the parameters of a PF.

An analytical solution to the problem of finding the maximum value of the log likelihood is in most practical situations difficult, if not impossible, to accomplish. However, we may use a search algorithm to find the maximum value of the log likelihood, and the corresponding values of a and b to any desired finite degree of precision. For example, the Nelder–Mead simplex method implemented in the *fminsearch* function in MATLAB is well-suited to this problem.

Given a large enough number of trials in an experiment, the likelihood function is typically unimodal, like that shown in Figure 4.13. It should be mentioned here that the results of the above-described and very modest experiment were not typical: most repetitions of the experiment would not result in unimodal likelihood functions. Whether an experiment will result in a unimodal likelihood function depends on other factors besides N. From our own experience these factors include the guess rate, the lapse rate, and the distribution of stimulus intensities used in the experiment.

Function **PAL_PFML_Fit** in the Palamedes toolbox estimates any or all of the four parameters associated with a PF using the maximum likelihood criterion. It utilizes the *fminsearch* function in MATLAB to maximize the log likelihood. Its usage is as follows:

```
[paramsValues LL exitflag output] = PAL_PFML_Fit(StimLevels, ...
NumPos, OutOfNum, paramsValues, paramsFree, PF)
```

The input variables are as follows:

StimLevels: vector containing the stimulus intensities utilized in the experiment. For the modest experiment described above we would define:

```
>>StimLevels = [-2 -1 0 1 2];
```

NumPos: vector of equal length to **StimLevels** containing, for each of the stimulus levels, the number of positive responses (e.g., "yes" or "correct") observed. Thus, with reference to the above experiment we define:

```
>>NumPos = [2 3 3 3 4];
```

OutOfNum: vector of equal length to **StimLevels** containing the number of trials tested at each of the stimulus levels. To fit the above experiment we define:

```
>>OutOfNum = [4 4 4 4 4];
```

paramsValues: vector containing values for each of the four parameters of the PF: alpha (threshold); beta (slope); gamma (guess rate); lambda (lapse rate). We need to provide a value for each of these parameters. For those we wish to fit we provide the initial search values (we would use our best guesses for the parameter values), for those parameters we wish to remain fixed we provide their assumed or known fixed values. Let us say that in the above experiment we wish to fit the threshold and the slope, but assume fixed values for the guess rate and the lapse rate. Our initial guess for the value of log alpha is 0, our initial guess for the slope is 1. Since data were obtained in a 2AFC experiment we fix the guess rate at 0.5, and since we have high confidence in our observer's vigilance, we set the lapse rate to 0. We define:

```
>>paramsValues = [0 1 .5 0];
```

paramsFree: vector containing a code for each of the four parameters indicating whether it is a free parameter (coded by 1) or a fixed parameter (coded by 0). As noted above, we wish to estimate the threshold and the slope, but we wish the guess rate and the lapse rate to be fixed parameters. We define:

```
>>paramsFree = [1 1 0 0];
```

PF: Of course, we need to specify the form of the PF we wish to fit. We will pass the appropriate function as a MATLAB *inline* function to the routine. PFs supported are those described above (Section 4.3.2): the Logistic (**PAL_Logistic**), Weibull (**PAL_Weibull**), Cumulative Normal (**PAL_CumulativeNormal**), Hyperbolic Secant (**PAL_HyperbolicSecant**), and Gumbel (**PAL_Gumbel**) functions. We will use the Logistic function and define:

```
>>PF = @PAL_Logistic;
```

We may now place our function call:

```
>>[paramsValues LL exitflag output] =...
PAL_PFML_Fit(StimLevels,NumPos, OutOfNum, ...
paramsValues, paramsFree, PF)
```

The output in MATLAB is as follows:

```
paramsValues =
0.0102   0.9787   0.5000   0
```

```
LL =
-10.1068
exitflag =
1
output =
iterations: 31
funcCount: 58
algorithm: 'Nelder-Mead simplex direct search'
message: [1 × 196 char]
```

paramsValues again contains our four PF parameters. For those parameters we wished to estimate (here, the slope and the threshold) **PAL_PFML_Fit** returns the best-fitting values for the threshold and slope parameters. For those parameters we indicated to be fixed, **PAL_PFML_Fit** simply returns the fixed values we chose (0.5 for the guess rate, and 0 for the lapse rate). The scale on the returned value of α (threshold) will be the same as the scale we used to define our stimulus levels (in **StimLevels**). Here, we provided the function with log-transformed values of our stimulus levels, and thus the first value in **paramsValues** returned by the function is actually $\log\alpha$.

LL is the value of the log likelihood (calculated as in Equation 4.12) associated with the fit.

exitflag is the exit flag of *fminsearch*. A value of 1 indicates that the search terminated successfully. Note that an exit flag equal to 1 does not necessarily mean that we found the global maximum likelihood; we may have found a local maximum. A solution that has converged on anything other than the global maximum is usually easily detected, since local maxima typically occur at parameter values that are clearly wrong. Comparison of our estimates with Figure 4.13 indicates that the obtained parameter estimates correspond to the global maximum likelihood.

output: the output structure provides information about the simplex search. It gives the number of iterations performed, the number of function evaluations performed, the algorithm used, and a message. When we type:

```
>>output.message
```

MATLAB tells us:

```
Optimization terminated:
the current x satisfies the termination criteria using
OPTIONS.TolX of 1.000000e-004
and F(X) satisfies the convergence criteria using OPTIONS.
TolFun of 1.000000e-004
```

The above message contains the utilized criterion for the tolerance on the estimates of the parameters (**OPTIONS.TolX**) and on the value of the log likelihood (**OPTIONS.TolFun**).

The values listed in **output.message** are the default values used in the *fminsearch* function in MATLAB, simply because we did not tell MATLAB to do otherwise. We can choose to perform the fitting using different criterion values for tolerance. We can also override other default settings. This requires us first to change the values in the options structure that holds these values. First we create an options structure by copying the default options structure:

```
>>options = optimset('fminsearch');
```

We can then change individual entries. For example:

```
>>options.TolFun = 1e-09;
```

We then pass the new options structure to function **PAL_PFML_Fit** by adding an extra argument **'SearchOptions'**, followed by the new **options** structure:

```
>>[paramsValues LL exitflag output] =...
PAL_PFML_Fit(StimLevels, NumPos, OutOfNum,paramsValues, ...
paramsFree,PF,'SearchOptions',options);
```

For additional information about the **options** structure type:

```
>>help optimset
```

or visit the MATLAB Function Reference section for the function **optimset**.

Another optional argument allows us to constrain the estimate of the lapse rate to a specified range of values. To do this, we add an argument **'LapseLimits'**, followed by a 1×2 vector containing the lower limit and the upper limit we wish to place on the lapse rate parameter estimate. For example, the call:

```
>>[paramsValues LL exitflag output] =...
PAL_PFML_Fit(StimLevels, NumPos, OutOfNum,paramsValues, ...
paramsFree,PF,'LapseLimits',[0 0.06]);
```

would constrain the lapse rate parameter to have a value between 0 and 0.06 (inclusive). Keep in mind that this optional argument only makes sense when the lapse parameter is set to be a free parameter and will be ignored otherwise.

It is not necessary to combine trials with equal stimulus intensities in single entries in our vectors **StimLevels**, **NumPos**, and **OutOfNum**. For example, it may be convenient to have as many entries in these vectors as we have trials in the experiment. In this case, we would obtain identical results if we define **StimLevels**, **NumPos**, and **OutOfNum** as follows:

```
>>StimLevels = [-2 -2 -2 -2 -1 -1 -1 -1 0 0 0 0 1 1 1 1 2 ...
2 2 2];
```

```
>>NumPos = [1 0 1 0 0 1 1 1 1 1 0 1 1 1 1 0 1 1 1 1];
>>OutOfNum = ones(1,20);
```

The m-file **PAL_PFML_Demo.m** demonstrates use of the function **PAL_PFML_Fit**.

4.3.3.1.3 Error Estimation

The above section describes how to find the best-fitting parameters of a PF using the maximum likelihood criterion. Because our estimates are based on what is necessarily a limited number of observations we realize, though, that our estimates are exactly that: estimates. The estimates we have obtained from our sample will not be exactly equal to the true parameter values. If we were to repeat our experiment under exactly identical conditions and estimate our parameter values again, our second set of estimates will also not be equal to the true parameters, nor will they be equal to our original estimates. This is simply due to noise; our observers have a noisy brain, our stimuli may have a stochastic element, etc.

So, we will arrive at different threshold estimates when we repeat an experiment, even when we use identical conditions. We call this sampling error. This is a problem, especially when we are interested in determining whether experimental manipulations affect the parameters of our PF. We can never know for sure whether differences between parameter estimates in different conditions are a result of the experimental manipulation or sampling error. However, there are methods available to us to make at least a reasonable guess whether the experimental manipulation affected the parameters of the PF. Here we will discuss a method which allows us to estimate by how much we can expect our parameter estimates to differ from the true value of the parameter.

So, we would like to gain some information as to the magnitude of the error to which our parameter estimates are subject. We will address this issue by considering a different, but similar question. The question we ask ourselves is: given particular values for the true parameters and an infinite number of experiments like ours, all resulting in estimates of the true parameters, what degree of error are these estimates subject to? More generally, given particular values for the true parameters, what are the probability density functions of our parameter values? This is, of course, a hypothetical question because we will never know the true values of the parameters, and we certainly do not wish to repeat our experiment an infinite number of times. Not only is this question hypothetical, but it also seems backward. When we have completed our single experiment and fit our data we will know the parameter estimates and would like to know the true parameter values, but the question assumes knowledge of the true parameters and asks about the parameter estimates. As backwards as this question may appear, let us attempt to answer it anyway, and see where that leads us.

The distribution of parameter estimates resulting from a hypothetical infinite number of experiments is known as a "sampling distribution." In certain circumstances sampling distributions for parameters may be derived analytically. These are typically algebraically simple parameters, such as the mean. As an example, let us consider the sampling distribution of the mean. By the well-known Central Limit Theorem (e.g., Hays, 1994), given a population of scores with mean μ and standard deviation σ, the sampling distribution of the (sample) mean $\bar{X}$ will have a mean ($\mu_{\bar{X}}$) equal to μ and a standard deviation ($\sigma_{\bar{X}}$) equal to $\sigma/\sqrt{N}$, where N is the sample size. Moreover, if our sample size N is large enough, the sampling distribution of the mean will closely approximate the normal distribution. By way of example, let us say that we have a population with known mean $\mu = 100$ and standard deviation $\sigma = 20$. If we were to collect an infinite number of samples of size $N = 100$, and for each of these samples calculated the sample mean $\bar{X}$, the resulting distribution of sample means would have mean $\mu_{\bar{X}}$ equal to 100 and the standard deviation of sample means (the standard error) would be equal to $\sigma_{\bar{X}} = \sigma/\sqrt{N} = 2$. If we use the normal distribution as an approximation to the sampling distribution (which is appropriate given that our $N = 100$), we are now in a position to make some interesting statements. For example, approximately 68% of our sample means would have a value between 98 and 102 $(\mu_{\bar{X}} \pm \sigma_{\bar{X}})$. This may, of course, be paraphrased as: 68% of sample means would be in error by less than 2 points. We can also determine that a very large proportion (>0.9999) of sample means would have a value between 92 and 108, which may be paraphrased as: almost all sample means would be in error by less than 8 points. We may paraphrase even further, and state that if we were to take a single sample of size $N = 100$ from this population there is a probability of about 68% that the sample mean will be in error by less than two points, and also that the probability that the sample mean will be in error by less than 8 points is near unity.

Let us now envision a situation in which we have a large population of scores and we wish to estimate the mean of this population. Somehow we know that the standard deviation of the population is equal to 20. We reason that if we take a sample of $N = 100$ and calculate the sample mean, the probability that this sample mean will be in error by less than 2 points is approximately 68%. Next, we actually do take a sample of $N = 100$, and we calculate the sample mean which comes out at, say, 50. It sounds as if we can now argue that there is a 68% probability that our sample mean is in error by less than 2 points and thus, that the probability that the population mean has a value between 48 and 52 is 68%. We cannot make this argument though, simply because the experiment is over and the sample mean is known, and it is either in error by less than 2 points or it is not. Similarly, the population mean is not a random variable. That is, it either has a value between 48 and 52 or it does not. Despite the fact that we cannot use the term probability, it seems like a reasonable argument. We present our statement using the term "confidence." We might say: "We are 68% *confident* that the population mean has a value between 48 and 52." Mission accomplished!

Thus, once we have conducted our psychophysical experiment, we would like to know the characteristics of the sampling distributions of our parameters. From this, we could determine by how much we can expect our estimate for, say, the threshold to be in error. To determine the sampling distribution, we would first need to know the population parameters, which of course we do not (if we did, there would be no need for our experiment). This problem also exists when we wish to estimate the mean of a distribution as we did above. Our example there assumed knowledge of the population standard deviation. In any practical situation, the population standard deviation will be just as unknown as the population mean we wish to estimate. The solution is to consider the sample to be representative of the population, and to use the sample to estimate the population parameters. In the case of the mean we would have to estimate the population standard deviation from our sample (our sampling distribution would then have to be the t-distribution, but that aside). We could do the same for our purposes here. Even though we do not know the true PFs parameters, we have estimates of them based on our sample. For the purposes of finding our sampling distribution we will use these as estimates of the true PFs parameters and derive a sampling distribution for a PF with these parameters.

Our second problem is that even if we assume our population parameters to equal our sample parameters, no one as of yet has analytically derived the sampling distribution of, say, the threshold parameter of a PF as estimated by the maximum likelihood method. However, we may approximate our sampling distribution by simulating our experiment many times using the estimated true PF parameters to define the generating function. That is, we simulate an observer to act according to our estimated true PF. We run this observer through simulations of our experiment many, many times. Each time we derive estimates of the PFs parameters. These estimates then serve as our empirically-derived sampling distribution.

An example is probably in order here. We wish to measure the threshold of a human observer on a 2AFC psychophysical task. We use the method of constant stimuli to collect 100 responses at each of 7 equally spaced (in log units) stimulus levels: −3, −2, . . . , 3. The observed number of correct responses at these stimulus levels are: 55, 55, 66, 75, 91, 94, and 97, respectively. We can use **PAL_PFML_Fit** to find the best-fitting PF. We wish to estimate the threshold and slope, but we fix the guess rate at 0.5 and we will also assume that the lapse rate equals 0. In order to perform the fit, we type:

```
>>PF = @PAL_Logistic;
>>StimLevels = [-3:1:3];
>>NumPos = [55 55 66 75 91 94 97];
>>OutOfNum = 100.*ones(size(NumPos));
>>paramsValues = [0 1 .5 0];
>>paramsFree = [1 1 0 0];
>>paramsValues = PAL_PFML_Fit(StimLevels,NumPos, ...
OutOfNum, paramsValues, paramsFree, PF)
```

The output in MATLAB is:

```
>>paramsValues =
-0.1713   0.9621   0.5000   0
```

Thus our best maximum likelihood estimate for the threshold is −0.1713 (in log units, since we entered **StimLevels** in log units), and for the slope is 0.9621. We now wish to obtain the sampling distribution of our parameter estimates. This will allow us to get some idea as to how much error our parameter estimates might be subject to. We imagine an observer whose PFs true parameters are those we have estimated from our observer (i.e., $\log \alpha = -0.1713$, $\beta = 0.9621$). We then simulate this observer as a participant in our experiment many, many times. In the simulations, we use the same stimulus intensity values and the same number of trials at each of the stimulus intensities as we did for our human observer. From the results of each simulated experiment, we estimate the PFs parameters. For all these estimates we know exactly by how much they are in error, because we know the true parameters of the generating PF. Our distributions of these estimates are our empirically derived sampling distributions. The left panels of Figure 4.14 show sampling distributions of the estimates for the threshold and slope based on B = 40,000 simulated experiments. The results of all these simulated experiments were generated by a PF with known parameter values ($\log \alpha = -0.1713$, $\beta = 0.9621$). These generating parameters are indicated in the figure by the vertical lines through the histograms.

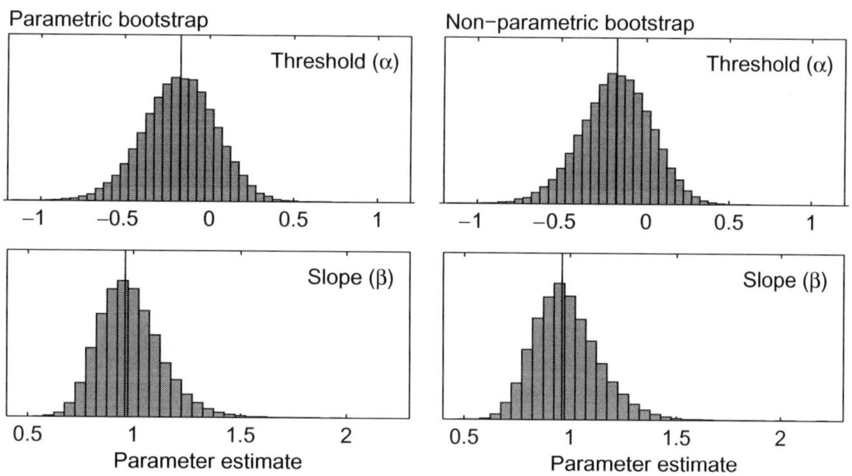

FIGURE 4.14 Empirical sampling distributions of the threshold and slope parameters of a PF. Distributions on the left were obtained using a parametric bootstrap, those on the right using a nonparametric bootstrap. Vertical lines indicate the best-fitting values of the threshold and slope of the results of the human observer, which are also the values used to generate the bootstrap samples.

Let us make a few observations. First, the mean of the sampling distribution for the threshold equals −0.1761. This is quite close in value to the threshold of the generating PF. This means that, although some estimates were too high and others too low, on average they were approximately on target; our fitting procedure results in threshold estimates that have little bias, at least under the conditions of this experiment. Second, the mean of the sampling distribution for the slope equals 0.9835. Again some estimates were too high, some were too low, but now the average, frankly, is a little high too (compare to generating slope 0.9621). Thus, our estimates for the slope parameter have, on average, overestimated the true slope parameter a bit. We also note that, whereas the sampling distribution of thresholds is symmetrical, that for slopes is positively skewed. In skewed distributions, the median is sometimes used as a measure of central tendency. The median of the sampling distribution for the slope equals 0.9705, closer to the true (i.e., generating) value, but still too high.

The bias of our estimates says, of course, nothing about how much any given *individual* estimate may be in error. For that we need to look at the standard deviation of our sampling distribution of parameter estimates (i.e., "the standard error of estimate" or SE). The SE is calculated as:

$$SE_{\hat{\alpha}} = \sqrt{\sum_{b=1}^{B}(\hat{\alpha}_b - \bar{\hat{\alpha}})^2 \Big/ (B-1)}, \tag{4.13}$$

where B equals the number of simulations (here, $B = 40{,}000$), $\hat{\alpha}_b$ is the threshold estimate resulting from bootstrap simulation b (actually, since we have used a log-transform of stimulus intensities, in this particular example we should use the log-transformed value of the threshold estimate), $\bar{\hat{\alpha}}$ is the mean threshold estimate (again, here, we should use the mean of the log-transformed threshold estimates).

The standard error of the threshold (calculated as in Equation 4.13) equals 0.2121. Since we note that the shape of the sampling distribution is approximately normal, this would mean that about 68% of threshold estimates deviated from the generating threshold by less than 0.2121 log units. We may assume that the true generating PF of our human observer is much like that used to create these sampling distributions (after all, we modeled our simulated observer after our human observer), and thus that the sampling distribution of our human observer's threshold would be much like that shown in the figure. Thus, we can now make statements such as: "We can be 68% confident that our human observer's true threshold has a value in the range -0.1713 ± 0.2121."

Calculated analogously to the standard error of the threshold, the standard error of the slope estimate is equal to 0.1461. The shape of this sampling distribution deviates systematically from normal, so we cannot translate this value easily into a confidence interval. Also, as the sampling distribution of the slope is asymmetrical, we might wish to determine two standard errors: one for estimates that were below the generating value; the other for estimates that were above the generating value. The SE for low estimates equals 0.1300, and that for the high estimates is 0.1627.

A good case can be made for an alternative calculation of standard errors. In Equation 4.13, we have used deviations from the sampling distribution's mean threshold $(\hat{\alpha}_b - \bar{\alpha})$ as our basis for the calculation of the SE. However, these values are not truly the errors of estimate. We know the true threshold that generated the sampling distribution, thus we need not estimate it. The true error of estimate of any $\hat{\alpha}_b$ is not $(\hat{\alpha}_b - \bar{\alpha})$, it is $(\hat{\alpha}_b - \alpha_g)$, where α_g is the threshold of the PF that generated the sampling distribution. If we use instead the equation:

$$SE_{\hat{\alpha}} = \sqrt{\sum_{b=1}^{B} (\hat{\alpha}_b - \alpha_g)^2 \Big/ B} \tag{4.14}$$

to calculate the standard error of estimate, we find that the value is nearly identical in value to that obtained using Equation 4.13. Note that the denominator now is B, since we use the known generating α_g and did not estimate it by $\bar{\hat{\alpha}}$. Whether we use $\bar{\hat{\alpha}}$ or α_g has very little effect on the obtained standard error of estimate; this is because $\bar{\hat{\alpha}}$ and α_g are nearly identical in value. However, when we make the same adjustment in the calculation of the standard error of estimate of the slope, we do arrive at somewhat different estimates (0.1477 versus 0.1461). The difference can be attributed to $\hat{\beta}$ having overestimated β_g. When we calculate separate estimates for the low estimates and the high estimates using β_g, we arrive at SEs of 0.1199 (at the lower end) and 0.1690 (at the higher end), compared to 0.1300 and 0.1627, respectively, when we use $\hat{\beta}$.

The procedure above is referred to as "parametric" bootstrapping. The sampling distribution was created using a simulated observer characterized by the parameters of the best-fitting PF to our human observer's data. In other words, we have assumed that our human observer's true PF is, in this case, the logistic. The accuracy of our obtained estimated standard errors relies on this assumption being correct.

We may also perform a "non-parametric" bootstrap procedure, in which we do not summarize our human observer by a parametric description of his or her assumed PF to generate the sampling distribution. Rather, we use the observed proportions correct at each of the stimulus intensities directly to generate the bootstrap simulations of the experiment. For example, here the human observer responded correctly on proportions 0.55, 0.55, 0.66, 0.75, 0.91, 0.94, and 0.97 of trials at each of the stimulus intensities, respectively. We may run our simulations without first summarizing this performance by a PF, as we did above. We may instead use these proportions directly in our simulations. That is, at each of the trials at stimulus intensity −3, our simulated observer will generate a correct response with probability 0.55, etc. The panels at the right hand side of Figure 4.14 display the sampling distributions of the threshold and slope generated by a non-parametric bootstrapping procedure. The SEs obtained are close in value to those obtained by the parametric procedure (0.2059 versus 0.2121 for the SE in threshold, and 0.1530 versus 0.1461 for the SE in slope). The two methods generate very similar results here, because our human observer's proportions correct are well described by our fitted PF. When our fitted PF is not a very good fit, the two bootstrapping methods might generate somewhat different results. Thus, it is good practice to perform a goodness-of-fit test of the PF and perform a parametric bootstrap only when the PF fits the data well. When the goodness-of-fit is unacceptable, one should perform a non-parametric bootstrap.

A few more words of caution should be given here. The accuracy of our SEs depends critically on the accuracy of our parameter estimates based on our human observer (e.g., Efron & Tibshirani, 1993; Kuss, Jäkel, & Wichmann, 2005). The SEs estimated by the parametric bootstrap are actually those that are associated with the function that we used to generate our bootstrap samples. They are accurate only to the extent that our human observer's PF corresponds to our estimate of it. We should realize that our estimated slope for the human observer's PF might be biased. When we use a positively biased estimate as the slope of the generating PF in our bootstrap simulations, the bootstrap simulations will be generated by a less noisy observer than our human observer. This will lead to an underestimation of the SE of the threshold parameter. For this reason we might prefer to perform the non-parametric bootstrap procedure, which does not involve estimating the human observer's threshold or slope in order to perform the simulations.

On the other hand, when we use an adaptive method during our experiment (Chapter 5), the parametric bootstrap procedure might be preferable. This is because, when we use an adaptive procedure, we may have very few observations at any utilized stimulus intensity. Consider a situation in which only one trial was presented at stimulus intensity x. Let us assume that our human observer produced a correct response on this trial. When we use the non-parametric bootstrap procedure in this case, our simulated observer will, in all of the simulations, also produce a correct response to this trial. In the extreme case where any given x was utilized on only one trial, the simulated observer will respond identically to our human observer on every trial in all simulations. In this case the parameter estimates from all simulations will, of course, be identical to the estimates of our human observer. Our SEs of estimate will then, of course, be zero.

The function **PAL_PFML_BootstrapParametric** in the Palamedes toolbox performs a parametric bootstrap simulation. Its syntax is as follows:

```
>>[SD paramsSim LLSim converged] =...
PAL_PFML_BootstrapParametric(StimLevels, OutOfNum, ...
paramsValues, paramsFree, B, PF)
```

The input arguments **StimLevels**, **OutOfNum**, **paramsValues**, **paramsFree**, and **PF** are as defined in the function **PAL_PFML_Fit**. For these we will not provide details again (please refer to Section 4.3.3.1.2), but instead demonstrate their usage by way of example below. The input argument **B** is the number of simulations to be performed in the bootstrap. The more simulations we include in our estimate of the standard error of estimate the more reliable it becomes, of course. The function will, however, also take longer to execute.

For the purposes of estimating the standard error of estimates, a few hundred simulations should suffice (e.g., Efron & Tibshirani, 1993). In order to replicate, albeit on a smaller scale ($B = 400$), the bootstrap simulation from above, we would type:

```
>>PF = @PAL_Logistic;
>>StimLevels = [-3:1:3];
>>OutOfNum = 100.*ones(size(StimLevels));
>>paramsValues = [-0.1713 0.9621 .5 0];
>>paramsFree = [1 1 0 0];
>>B = 400;
>>[SD paramsSim LLSim converged] = ...
PAL_PFML_BootstrapParametric(StimLevels, OutOfNum, ...
paramsValues, paramsFree, B, PF);
```

The function will simulate an observer whose PF is characterized by **PF** and **paramsValues**, and run this observer through the experiment for a total of **B** simulations. The stimulus intensities used and the number of trials at each of these intensities are those specified by **StimLevels** and **OutOfNum**, respectively. The results of each simulation will be fit with a PF of type **PF**. After the function finishes, we may inspect the standard error of estimates:

```
>>SD
SD =
0.2119   0.1585   0   0
```

The first two entries are the estimates of the standard error of estimate for the threshold and slope as calculated by Equation 4.13. These values will vary somewhat each time we perform a bootstrap, due to the stochastic nature of the bootstrap procedure. We note, though, that those obtained here are in close agreement to those obtained above. Since we specified (in **paramsFree**) that the guess rate and lapse rate parameters are fixed parameters, these were not estimated for any of the simulations, and thus their standard error is zero.

The functions also returns (in **paramsSim**) the parameter estimates for each individual simulation. **paramsSim** is a **B** $\times$ 4 (alpha, beta, gamma, lambda) matrix. This may be useful in case one wishes to inspect characteristics of the sampling distributions. One may wish to check these distributions for symmetry, or for outliers, for example. Outliers may be indicative of a fit having found a local maximum in the likelihood function rather than the global maximum likelihood. **paramsSim** may also be used when one wishes to calculate SEs in a manner different from that of Equation 4.13. **LLSim** is a vector of length **B**, which contains for each of the bootstrap simulations the log likelihood associated with the fit. Finally, **converged** is a vector of length **B** which contains, for each simulated experiment, a 1 in case the simulated experiment was fit successfully or a 0 in case it was not. Of course, if this vector contains 0s, our sampling distributions are flawed. The function will issue a warning whenever a simulated experiment could not be fit successfully, for example:

```
Warning: Fit to simulation 226 of 400 did not converge.
>> In PAL_PFML_BootstrapParametric at 165
```

If **converged** contains zeros when all simulations have completed, the function will again issue a warning and indicate how many of the simulations successfully converged, for example:

```
Warning: Only 398 of 400 simulations converged
>> In PAL_PFML_BootstrapParametric at 173
```

When not all simulations result in a successful fit, there is no solution which is truly elegant; a few tempting, but inappropriate, ideas spring to mind. One idea would be to ignore the failed simulations and calculate the standard error across the subset of simulations that did converge. Another tempting, but equally inappropriate, idea would be to generate new simulated datasets to replace those which resulted in failed fits in the original set. One more seemingly innocent, but once again inappropriate, solution would be to try the entire set of, say, 400 simulations again. We could continue to produce sets of 400 simulations until we have a set for which all 400 fits were succesful. We would then calculate the standard errors across that complete set. The problem with all these ideas is that the resulting error estimates would not be based on a random sample from the population of possible simulations. Rather, the estimates would be based on a select subset of the population of simulations, namely those that can be fitted successfully.

So, what is one to do when some simulations fail to converge? Generally, fits are more likely to converge when we have fewer free parameters in our model. Thus, we could fix one of the free parameters. Another manner in which to increase our chances of having all simulations fit successfully is to gather more responses, because the chances of obtaining a succesful fit generally increase with an increasing number of responses. One final solution which requires that all but a very few fits converged succesfully is to calculate standard errors across the succesful fits only, but to acknowledge to our audience (and ourselves) that our error estimate is based on a sample which was not entirely random.

In order to perform a non-parametric bootstrap we use the function **PAL_PFML_BootstrapNonParametric**. This function is called as:

```
>>[SD paramsSim LLSim exitflag] =...
PAL_PFML_BootstrapNonParametric(StimLevels, NumPos, ...
OutOfNum, paramsValues, paramsFree, B, PF);
```

The function is similar to **PAL_PFML_BootstrapParametric**, except that it characterizes the simulated observer not by the best-fitting PF to our human observer's data, but rather uses the raw observed proportions correct at each of the stimulus intensities directly. All arguments in the function call are identical to those for **PAL_PFML_BootstrapParametric**, except that we now also provide the function with the number of correct responses at each of the stimulus intensities (**NumPos**):

```
>>NumPos = [55 55 66 75 91 94 97];
```

On completion of the function call, we may inspect the standard errors of estimate:

```
>>SD
SD =
0.2082   0.1576   0   0
```

Once again, these values closely match those obtained above. Return arguments **paramsSim**, **LLSim**, and **converged** are as above.

Note that here we also need not group our trials by stimulus intensity; if it is more convenient to pass **StimLevels**, **NumPos**, and **OutOfNum** such that every entry in these vectors corresponds to a single trial, we may do so. The function will group trials presented at identical stimulus intensities before it will calculate the proportions correct at each of these stimulus intensities. For reasons mentioned above, a warning will be issued when one or more of the stimulus intensities were used only on a single trial. The warning will be accompanied by a listing of all stimulus intensities used and the corresponding number of trials that were presented at these stimulus intensities.

Both **PAL_PFML_BootstrapParametric** and **PAL_PFML_BootstrapNonParametric** have a few optional arguments. The first is **'SearchOptions'** followed by an options structure. The use of this optional argument is analogous to its use in **PAL_PFML_Fit**, except that when we include the options structure in the call here its entries will be used when the simulated data sets are fitted. When we do not include the options structure, the default values will be used. A second optional argument allows for the function to retry failed fits. Some fits to simulated data may fail to converge, for example when the search algorithm gets stuck in a local maximum, or when it wanders into a region of parameter space which is not in the domain of the parameter values. Failed fits provide a serious nuisance for reasons mentioned above. Sometimes such failed fits can be salvaged by retrying the fit but starting the search at different parameter values. The initial search will start at the values specified in **paramsValues**. There is an option to have the function try failed fits again for any specified maximum number of tries, each time starting the search at different parameter values. In order to use this option we first need to set the value of **maxTries** to the maximum number of times we wish the function to try the fit. The default value of **maxTries** is 1. In other words, unless we change the value, the function will try each fit just once, and in the case where that initial fit fails it will give up. We can change the value of **maxTries** by passing the argument **'maxTries'** to the bootstrap function, followed by an integer argument indicating the maximum number of tries we would like the function to perform before it gives up. Each try the function will choose a new random starting value for the search. It is wise, but optional, to provide the function with a sensible range of starting values to pick from. This range is specified in the vector **rangeTries**, which has four entries, one for each parameter of the PF. For each parameter, starting values will be from a range of values as wide as specified in **rangeTries**, and centered on the value specified in **paramsValues**. For example, let's say that **paramsValues** is defined as **[2 50 .5 0]** and **RangeTries** is defined as **[2 60 0 0]**. The initial search will start with a threshold value of 2, and a slope value of 50. In case it fails, each of the following tries will start with a random value between 1 and 3 for the threshold and a random value between 20 and 80 for the slope. Each try will use a value of 0.5 for the guess rate and 0 for the lapse rate. Note that some simulated datasets may never be successfully fit, no matter what value you set **maxTries** to. As mentioned above, when this happens the routine will issue a warning. This may happen especially when the number of trials in an experiment is small, the number of free parameters is high and/or the stimulus levels are inappropriately chosen.

Finally, we have the option to constrain the lapse parameter to have a value within any range of values. We use the optional argument **'lapseLimits'**, followed by a vector containing the lower and upper limit of the interval as we did above in **PAL_PFML_Fit**.

The following is an example function call where all the optional arguments are used. It is a variation of the example above, but now we set the generating lapse parameter to 0.02 (note that in real life it would be inappropriate to use parameters that are different from those estimated or assumed in the fit to the observer's data). We free the lapse parameter in the simulated fits, but constrain it to have a value between 0 and 0.06:

```
>>PF = @PAL_Logistic;
>>StimLevels = [-3:1:3];
>>OutOfNum = 100.*ones(size(StimLevels));
>>paramsValues = [-0.1713 0.9621 .5 0.02];
>>paramsFree = [1 1 0 1];
>>B = 400;
>>options = optimset('fminsearch');
>>options.TolFun = 1e-09;
>>[SD paramsSim LLSim converged] =...
PAL_PFML_BootstrapParametric(StimLevels, OutOfNum, ...
paramsValues, paramsFree, B, PF, 'SearchOptions', ...
options, 'maxTries', 10, 'rangeTries', [.2 1.9 0 0.04], ...
'lapseLimits',[0 .06]);
```

Note that with the lapse parameter freed, some fits might fail initially but succeed on a successive try. In the example above, you would be able to tell whether a fit has failed on the first try when MATLAB produces a message such as:

```
<<Exiting: Maximum number of function evaluations has been
exceeded
- increase MaxFunEvals option.
Current function value: 349.536759
```

To prevent such messages, set the **Display** field in the **options** structure you pass to the function to **'off'**. By default, the bootstrap functions in Palamedes do this, but in the example above we have overridden this default by passing an **options** structure to the function in which **Display** is set to **'on'** (this is the default value MATLAB assigns when **optimset** is called).

4.3.3.2 Bayesian Criterion

4.3.3.2.1 Bayes' Theorem

The likelihood associated with assumed values a and b for the threshold and slope, respectively, of the PF is equivalent in value to the probability that a PF with $\alpha = a$ and $\beta = b$ would result in the exact outcome of the experiment as we have already observed it. As discussed above (Section 4.3.3.1), this likelihood can be interpreted neither as the probability of our exact experimental outcome having occurred, nor as the probability that the threshold has value a, and the slope has value b. A somewhat similar issue exists in classical ("Fisherian" or "frequentist") Null Hypothesis testing. The p-value that is the inevitable final result of a classical hypothesis test, and which eventually leads us either to reject the Null Hypothesis (if $p < 0.05$ or some other criterion value) or accept it (otherwise) is, as our statistics instructors have stressed to us, not the probability that the Null Hypothesis is true. Rather, the p-value we calculate in a classical hypothesis test corresponds to (something like) the probability that our experimental results could have occurred by sheer coincidence if, in fact, the null hypothesis were true. We can write this probability as $p(D \mid H)$, where D represents the outcome of our experiment (D stands for "data") and H is shorthand for "the Null Hypothesis is true." Of course, in classical testing, we do not consider D to be the exact outcome of our experiment, rather we consider D to be a range of outcomes (specified before we started collecting our results) that are unlikely to be obtained if H were true. Notwithstanding the tremendous intuitive appeal of the validity of concluding that H is likely false if $p(D \mid H)$ is small, this conclusion is nevertheless without merit. To conclude that H is unlikely given our experimental results is to make a statement regarding the value of $p(H \mid D)$; unfortunately, we have a value only for $p(D \mid H)$. However, we may relate $p(H \mid D)$ and $p(D \mid H)$ using Bayes' Theorem:

$$p(H \mid D) = \frac{p(H)p(D \mid H)}{p(H)p(D \mid H) + p(\overline{H})p(D \mid \overline{H})} = \frac{p(H)p(D \mid H)}{p(D)} \tag{4.15}$$

One may think of Equation 4.15 as expressing the central Bayesian concept that we use our experimental results D as serving to adjust the probability of H as we estimated it before considering the results of our experiment. As an illustration of this concept let us consider the entirely hypothetical case of Mr J. Doe. As part of his routine annual medical exam, Mr Doe is administered a diagnostic test D which tests for the presence of the rare medical condition H. Test D is known to be highly accurate; whereas the test results will be positive (D^+) for 99% of those individuals afflicted with the condition (H^+), test results will be negative (D^-) for 99% of those individuals not afflicted with the condition (H^-). We may write this as $p(D^+ \mid H^+) = 0.99$ and $p(D^+ \mid H^-) = 0.01$. In other words, test D diagnoses 99% of individuals correctly, whether they are afflicted with H or not.

Unfortunately, Mr Doe's test results are positive. Applying Fisherian logic leads us to conclude that Mr Doe is afflicted with H. After all, the probability that Mr Doe would test positive under the hypothesis that he is not afflicted with H is quite low: $p(D^+ \mid H^-) = 0.01$.

Mr Doe's outlook is not as bleak, however, when considered from a Bayesian perspective. The Bayesian perspective considers, besides the test results, another piece of information; if you will recall, medical condition H is rare. Let us assume that it is known that a proportion of only 1/10,000 of the population is afflicted with H. We may write this as $p(H^+) = 0.0001$. $p(H^+)$ is known as the "prior probability" of H^+. That is, prior to learning of Mr Doe's test results, our best estimate for the probability that Mr Doe was afflicted with H would have been 0.0001. In the Bayesian framework, the positive test result is considered to be merely a second piece of evidence which is used to adjust our prior probability, now also taking into account Mr Doe's positive test result, to derive the posterior probability $(H^+ \mid D^+)$. According to Bayes' Theorem:

$$
\begin{aligned}
p(H^+ \mid D^+) &= \frac{p(H^+)p(D^+ \mid H^+)}{p(H^+)p(D^+ \mid H^+) + p(H-)p(D^+ \mid H^-)} \\
&= \frac{0.0001 \cdot 0.99}{0.0001 \cdot 0.99 + 0.9999 \cdot 0.01} \\
&= \frac{0.000099}{0.000099 + 0.009999} \\
&\approx 0.0098
\end{aligned}
$$

In words, despite Mr Doe's positive test result, the odds are still strongly in favor of Mr Doe not being afflicted with H.

The obtained value for the posterior probability flies in the face of common sense. Indeed, students introduced to Bayesian reasoning by way of the above example often suspect something akin to sleight-of-hand, even if these students agree with every intermediate step performed. MDs, somewhat disconcertingly, do not do well either when it comes to Bayesian reasoning (e.g., Hoffrage & Gigerenzer, 1998).

4.3.3.2.2 Bayes' Theorem Applied to the Likelihood $L(a, b \mid \mathbf{y})$

We might apply Bayes' theorem to derive the posterior probability density function on our values for a and b. The situation is a little different here, since the likelihood is a function of a and b, which are continuous variables. In practice, however, we discretize our likelihood function. The appropriate formulation of Bayes' Theorem in this case becomes:

$$p(a,b \mid \mathbf{y}) = \frac{L(a,b \mid \mathbf{y})p(a,b)}{\sum_a \sum_b L(a,b \mid \mathbf{y})p(a,b)} \tag{4.16}$$

where $L(a, b \mid \mathbf{y})$ is our likelihood function as calculated in Section 4.3.3.1 and $p(a, b)$ is the prior distribution. The resulting posterior distribution $p(a, b \mid \mathbf{y})$ is a probability density function. That is, it allows one to determine the probability that the values of a and b lie within a specified range of values.

What should we use as our prior distribution? The prior distribution should, according to the Bayesian framework, reflect our prior beliefs regarding the values of the threshold and slope of our PF. We might perhaps base our prior beliefs on research preceding ours. We might also base our prior beliefs on informal pilot experiments. Defining a prior is, of course, somewhat of a subjective exercise. For this reason, and as one might imagine, the Bayesian approach is not without its critics.

Before we will argue that the Bayesian approach does not need to be as subjective an exercise as one may have concluded from the above, let us first illustrate Equation 4.16 by example. Let us imagine that, prior to performing our experiment of which Figure 4.13 shows the likelihood function, we formed beliefs regarding the values of the threshold and slope. Perhaps we did so by considering existing literature on similar experiments, or perhaps we did so based on our own informal pilot experiments. Either way, let us imagine that we judge our prior beliefs to be well described by the 2D Gaussian:

$$p(a,b) = \frac{1}{2\pi\sigma^2}\exp\left(-\frac{(\log a)^2 + (\log b)^2}{2\sigma^2}\right)$$

with $\sigma = 0.5$. This prior distribution is illustrated in Figure 4.15. Also shown in Figure 4.15 is our likelihood function again, as well as the posterior distribution derived by Equation 4.16. It is clear that the posterior distribution is determined primarily by our choice of the prior, and bears little resemblance to the likelihood function which was derived from our experimental data. We should keep in mind though, that our example is not very typical for two reasons. First, our prior beliefs are quite specific. For example, our prior indicates a belief that the probability that $\log \alpha$ has a value in the interval $(-0.5, 0.5)$ is near unity. Thus, apparently we already know a considerable amount about the values of α and β before we started

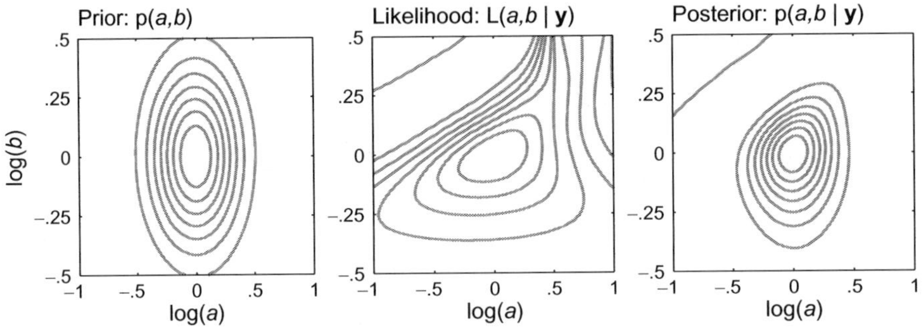

FIGURE 4.15 Contour plots of the prior distribution (left), the Likelihood function (middle), and the posterior distribution (right). The prior distribution reflects the researcher's beliefs regarding the value of the threshold and slope parameter before the results of the experiment are taken into account, the Likelihood function is based on the results of the experiment only, and the posterior distribution combines the prior and the Likelihood according to Equation 4.16.

our experiment. Second, our experiment was based on a very small number of trials ($N = 20$). Thus, it is not surprising that the results of our very small experiment hardly changed our strong prior beliefs.

For illustrative purposes, let us briefly consider two more priors, each at one of the two extreme ends of specificity. First, let us assume that the exact values of α and β are known with certainty before we conduct our experiment. Thus, our prior will be the unit impulse function located at the known values of α and β. Providing additional evidence by our experimental results will not alter our beliefs, and indeed our posterior will also be the unit impulse function located at the known values of α and β regardless of what our likelihood function might be. The prior at the other extreme of specificity is the uniform prior. The uniform prior does not favor any values of α and β over any others, and is thus consistent with a complete lack of knowledge or belief of what the values of α and β might be before we start our experiment. For this reason, some refer to the uniform prior as the "prior of ignorance." If our prior is the uniform prior, our posterior distribution is proportional to our likelihood function, and our parameter estimates will be determined entirely by the results of our experiment.

In practice it is difficult, if not impossible, to derive the continuous likelihood function analytically. Instead, we approximate the likelihood function across a discretized parameter space which is necessarily of finite extent. Thus, given that we can only consider a limited extent of the parameter space, a strictly uniform prior is not possible. The best we can do is to create a rectangular prior which, in essence, assigns a prior probability of 0 to values of α and β that lie outside our limited parameter space, but which, within the considered parameter space, favor no values of α and β over other values.

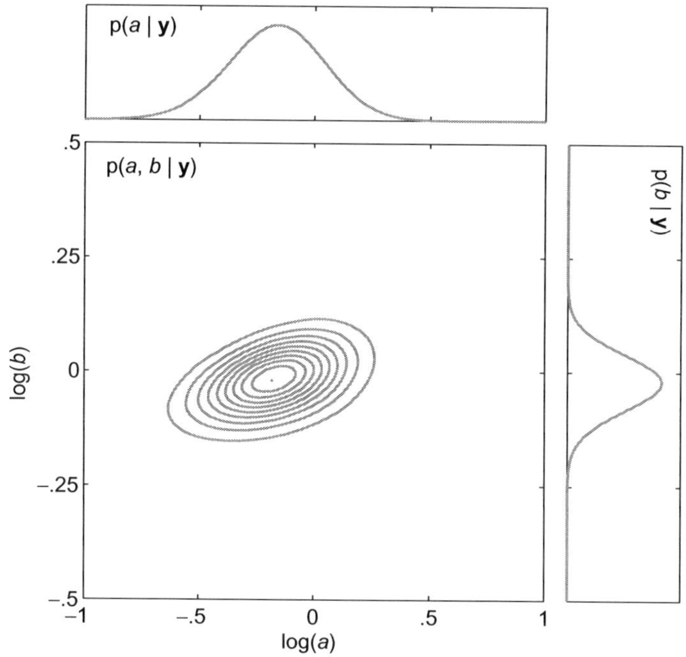

FIGURE 4.16 The posterior distribution.

We are finally ready to discuss how to derive parameter estimates based on the posterior distribution, and we will do so by example. Figure 4.16 presents the posterior distribution based on the 2AFC experiment we fitted using a maximum likelihood criterion before (Section 4.3.3.1). The experiment consisted of 700 trials, 100 trials at each of 7 stimulus intensities x which were equally spaced on a logarithmic scale between $\log(x) = -3$ and $\log(x) = 3$. The number of correct responses (out of the 100 trials) for each of the 7 stimulus levels was respectively: 55, 55, 66, 75, 91, 94, and 97. The prior used here was the uniform prior limited to $a \in [-1, 1]$, $b \in [-0.5, 0.5]$. We first calculated the likelihood function across the (discretized) parameter range defined by the prior (see Section 4.3.3.1). We used a guess rate (γ) equal to 0.5, and a lapse rate (λ) equal to 0. Since our prior is uniform and the posterior distribution is, by definition, a probability density function, our calculations simplify to:

$$p(a,b\,|\,\mathbf{y}) = \frac{L(a,b\,|\,\mathbf{y})}{\sum\limits_{a}\sum\limits_{b} L(a,b\,|\,\mathbf{y})} \tag{4.17}$$

That is, our posterior distribution is simply our likelihood function rescaled such that $\sum_a \sum_b p(a,b\,|\,\mathbf{y}) = 1$, which is a quality of any probability density function.

Also shown in Figure 4.16 are the marginal probability densities across a and b individually which are derived from $p(a, b \mid \mathbf{y})$ as follows:

$$p(a \mid \mathbf{y}) = \sum_b p(a, b \mid \mathbf{y}) \tag{4.18}$$

$$p(b \mid \mathbf{y}) = \sum_a p(a, b \mid \mathbf{y}) \tag{4.19}$$

Our Bayesian estimator of $\log \alpha$ is the expected value of $\log a$. That is:

$$\log \hat{\alpha} = E(\log a) = \sum_a \log a \; p(\log a \mid \mathbf{y}) \tag{4.20}$$

Similarly,

$$\log \hat{\beta} = E(\log b) = \sum_b \log b \; p(\log b \mid \mathbf{y}) \tag{4.21}$$

In this example, $\log \hat{\alpha} = -0.1715$ and $\log \hat{\beta} = -0.0225$.

We may note from Figure 4.16 that the parameter space included in the prior was such that it excluded only parameter values associated with extremely small likelihoods (remember that, using a uniform prior, the likelihood function is proportional to the posterior distribution, and thus Figure 4.16 may also be regarded as a contour plot of the likelihood function). As such, our exact choice for the range of values for $\log a$ and $\log b$ to be included in the prior would have a negligible effect on our final parameter estimates. Had we instead limited our prior to, for example, $a \in [-0.5, 0.5]$, $b \in [-0.25, 0.25]$ our likelihood function would have "run off the edge" of our prior (specifically the edge $\log a = -0.5$) and this would have affected our parameter estimates significantly. However, whenever we utilize a uniform prior which encompasses all but the extremely small likelihoods, the contribution of our subjective prior to our parameter estimates will be negligible.

Function **PAL_PFBA_Fit** derives the Bayesian estimators for the threshold and slope of a PF. A uniform prior across a limited parameter space, defined by the user, is used. Its syntax is as follows:

```
>>[paramsValues posterior] = PAL_PFBA_Fit(StimLevels, ...
NumPos, OutOfNum, priorAlphaValues, priorBetaValues, ...
gamma, lambda, PF)
```

The input variables are as follows:

StimLevels: vector containing the stimulus intensities utilized in the experiment. For the experiment described above we would define:

```
>>StimLevels = [-3:1:3];
```

NumPos: vector of equal length to **StimLevels** containing for each of the stimulus levels the number of positive responses (e.g., "yes" or "correct") observed. Thus, with reference to the above experiment we define:

```
>>NumPos = [55 55 66 75 91 94 97];
```

OutOfNum: vector of equal length to **StimLevels** containing the number of trials tested at each of the stimulus levels. To fit the above experiment we define:

```
>>OutOfNum = [100 100 100 100 100 100 100];
```

priorAlphaValues: vector of any length which specifies which threshold values a are to be included in the prior. Since our stimulus levels were defined in logarithmic units, we do the same for **priorAlphaValues**:

```
>>priorAlphaValues = [-1:.01:1];
```

priorBetaValues: vector of any length which specifies which slope values b are to be included in the prior. Values for b are defined in logarithmic units:

```
>>priorBetaValues = [-.5:.01:.5];
```

gamma: scalar corresponding to the assumed guess rate. Since the experiment is a 2AFC, we set **gamma** to equal 0.5:

```
>>gamma = 0.5;
```

lambda: scalar corresponding to the assumed lapse rate. We set **lambda** to equal 0:

```
>>lambda = 0;
```

PF: The psychometric function to be fitted. This needs to be passed as an inline function. We choose the Logistic function:

```
>>PF = @PAL_Logistic;
```

We are now ready to call our function:

```
>>[paramsValues posterior] = PAL_PFBA_Fit(StimLevels, ...
NumPos, OutOfNum, priorAlphaValues, priorBetaValues, ...
gamma, lambda, PF);
```

We suppress the output in MATLAB by following the command by a semicolon, because the output would include the entire matrix containing the posterior. The posterior is defined across the same parameter space as the prior and has 101 (length of **priorBetaValues** vector) $\times$201 (length of **priorAlphaValues**) entries in this example. However, we can inspect the parameter estimates by typing:

```
>>paramsValues
```

The output in MATLAB is:

```
paramsValues =
-0.1715   -0.0225   0.2106   0.0629
```

The first value corresponds to the Bayesian estimate of the log threshold (since we defined **StimLevels** and our prior in logarithmic units). The second value corresponds to the Bayesian estimate of the log slope. These values correspond quite closely to those derived using a maximum likelihood criterion (Section 4.3.3.1). If you will recall, the maximum likelihood estimate of log threshold was -0.1713, compared to -0.1715 using the Bayesian estimate here. That of the slope was 0.9621, compared to $10^{-0.0225} = 0.9495$ obtained here.

We may inspect the posterior distribution visually by typing, for example:

```
>>contour(posterior);
```

which produces a contour plot similar to that in Figure 4.16 above. Generally, it is a good idea to inspect the posterior distribution visually to check for edge effects. If the posterior is cut off abruptly at one or more of the edges of the parameter space, our particular choice of parameter space to be included in the prior distribution has a significant effect on our parameter estimates. In the example here, the contour plot of the posterior distribution indicates that edge effects will be negligible.

During the execution of **PAL_PFBA_Fit**, MATLAB may generate one or more **'Log of Zero'** warnings. These will occur whenever a likelihood associated with a particular stimulus intensity, response, and combination of a and b values is smaller than the smallest positive value that a double data type can represent (2.22507×10^{-308}). This might occur, for example, when we include in our prior distribution PFs with excessively steep slopes (i.e., high values for log b). The probability that a (hypothetical) observer characterized by a PF with an excessively steep slope would generate an incorrect response to a stimulus of high intensity is indeed near zero. In this situation, MATLAB assigns the value of zero to the likelihood, in essence excluding the PFs with the excessively steep slopes. These warnings may be disregarded. When we run the above example exactly as above, except that we now include excessively high values for the slope in the prior, for example:

```
>>priorBetaValues = [-.5:.01:2];
```

the **'Log of Zero'** warning is issued four times, but the values for our parameter estimates are not affected.

The user has the option to define a custom prior in case the uniform prior is deemed inappropriate. **PAL_PFBA_Fit** has an optional argument **'prior.'** The m-file **PAL_PFBA_Demo** demonstrates the use of this optional argument. If the **'prior'** argument is left out (as above) the uniform prior is used by default.

4.3.3.2.3 Error Estimation

The posterior distribution is a probability density function across the parameter space. As such, we may use it to derive probabilities of either parameter having a value within any range of values. For example, we may derive the probability that our (log-transformed) threshold parameter has a value between −0.25 and 0 as:

$$p(-0.25 < \log\alpha < 0) = \sum_{\log a \in (-0.25,0)} \sum_{\log b} p(\log a, \log b \mid \mathbf{y}) = 0.4267$$

Due to the discretization of the parameter space, this value will only be approximate. The finer the grid of our parameter space, the more accurate the approximation will be. We may derive the standard errors of estimate of our parameter estimates as the standard deviation of the posterior distributions:

$$SE_{\log\alpha} = \sqrt{\sum_a \sum_b (\log a - \log\hat{\alpha})^2 p(\log a, \log b \mid \mathbf{y})} \quad (4.22)$$

The SE for the threshold and slope are returned by the function **PAL_PFBA_Fit** as the third and fourth entry in the vector **paramsValues**, respectively. If you will recall, MATLAB returned **paramsValues** above as:

```
paramsValues =
-0.1715    -0.0225    0.2106    0.0629
```

As described above, the first and second entries (i.e., −0.1715 and −0.0225) are the Bayesian estimates of the log-transformed threshold value and slope, respectively. The third entry (0.2106) is the standard error on log threshold calculated as in Equation 4.22; the fourth entry (0.0629) is the standard error on log slope. The function also returns the entire posterior distribution such that we may choose to calculate our SEs in an alternative fashion. For example, we may calculate a lower SE and a higher SE if our posterior distribution is asymmetric. We note that the standard error on log threshold obtained here corresponds closely to that obtained from bootstrap analysis under the maximum likelihood framework (Section 4.3.31). There we estimated SE on log threshold as 0.2121 and 0.2059 using a parametric and a non-parametric bootstrap, respectively. The comparison of SE on slope is not quite as obvious, since here the value is the SE on log slope, whereas in the maximum likelihood bootstrap we determined the SE on the slope parameter proper. However, we note that $\log\beta + SE_{\log\hat{\beta}} = -0.0225 + 0.0629 = 0.0404$, which corresponds to a slope of $10^{0.0404} = 1.0975$, and that $\log_{\hat{\beta}} - SE_{\log\hat{\beta}} = -0.0854$, which corresponds to a slope of 0.8215. For comparison, under the maximum likelihood framework, we found that $\hat{\beta} = 0.9621$ and that $SE_{\hat{\beta}} = 0.1461$ using a parametric bootstrap and $SE_{\hat{\beta}} = 0.1530$ using a non-parametric bootstrap. Thus, $\hat{\beta} + SE_{\hat{\beta}}$ was $0.9621 + 0.1461 = 1.1082$ (parametric bootstrap) and $0.9621 + 0.1530 = 1.1151$ (non-parametric bootstrap), both of which correspond closely to the Bayesian analog obtained here (1.0975). We leave it to the reader to verify that the same is true of the values corresponding to 1 SE below our estimates.

Further Reading

Swets (1961) provides a very readable discussion of threshold theories. Chapter 6 of this text contains much more information on signal detection theory. Maximum likelihood estimation is a standard technique, and is discussed in any introductory statistical text. An excellent text on bootstrap methods is Efron and Tibshirani (1993). Our example of Mr Doe was adapted from many similar examples given in many places, among which is Cohen (1994) which discusses in a very readable manner some hypothesis testing issues.

Exercises

1. A two-alternative forced-choice experiment is conducted in which the log stimulus levels are −2, −1, 0, 1, and 2. 100 trials are presented at each of the stimulus levels. The observer responds correctly on respectively 48, 53, 55, 100, and 100 trials.
   a. Plot the results of this experiment
   b. By visual inspection, what do you estimate the 75% correct threshold to be?
   c. Use **`PAL_PFML_Fit`** to fit these data.
   d. Offer some suggestions to help improve on the design of the experiment.
2. As more trials are included in the computation of the likelihood associated with a parameter value, will this likelihood increase or decrease or does it depend on the outcome of the trial? Why?
3. It is said sometimes that "extraordinary claims require extraordinary evidence" (Carl Sagan coined the phrase). Explain how this statement relates to Bayes' theorem.

References

Cohen, J. (1994). The earth is round ($p < 0.05$). *American Psychologist*, *49*, 997–1003.

Efron, B., & Tibshirani, R. J. (1993). An introduction to the bootstrap. Boca Raton, FL: Chapman & Hall/CRC.

García-Pérez, M. A., & Alcalá-Quintana, R. (2005). Sampling plans for fitting the psychometric function. *The Spanish Journal of Psychology*, *8*(2), 256–289.

Green, D. M., & Swets, J. A. (1966). Signal detection theory and psychophysics. New York, NY: Wiley.

Hays, W. L. (1994). *Statistics*. Belmont, CA: Wadsworth Group/Thomson Learning.

Hoel, P. G., Port, S. C., & Stone, C. J. (1971). Introduction to Statistical Theory. Boston, MA: Houghton Mifflin Company.

Hoffrage, U., & Gigerenzer, G. (1998). Using natural frequencies to improve diagnostic inferences. *Academic Medicine*, *73*, 538–540.

Kuss, M., Jäkel, F., & Wichmann, F. A. (2005). Bayesian inference for psychometric functions. *Journal of Vision*, *5*, 478–492.

Meese, T. S., & Harris, M. G. H. (2001). Independent detectors for expansion and rotation, and for orthogonal components of deformation. *Perception*, *30*, 1189–1202.

Nachmias, J. (1981). On the psychometric function for contrast detection. *Vision Research, 21*, 215–223.

Quick, R. F. (1974). A vector-magnitude model of contrast detection. *Kybernetik, 16*, 65–67.

Swets, J. A. (1961). Is there a sensory threshold? *Science, 134*, 168–177.

Treutwein, B. (1995). Adaptive psychophysical procedures. *Vision Research, 35*, 2503–2522.

Wichmann, F. A., & Hill, N. J. (2001). The psychometric function: I. Fitting, sampling, and goodness of fit. *Perception & Psychophysics, 63*, 1293–1313.

CHAPTER

5

Adaptive Methods

5.1 INTRODUCTION

Measuring performance on a psychophysical task can be a time-consuming and tedious exercise. The purpose of adaptive methods is to make life easier for both observer and experimenter by increasing the efficiency of the testing procedure. Efficiency increases as the amount of effort required to reach a particular level of precision in the estimate of a parameter, such as the threshold of a psychometric function, decreases. Taylor and Creelman (1967) proposed a quantification of efficiency in the form of what they called the "sweat factor," which is symbolized as K and is calculated as the number of trials multiplied by the variance of the parameter estimate (the variance of the parameter estimate is, of course, simply the standard error squared). Adaptive methods aim to increase efficiency by presenting stimuli at stimulus levels where one might expect to gain the most information about the parameter (or parameters) of interest. Adaptive methods are so termed because they adjust the stimulus level to be used on each trial based on the responses to previous trials.

Many specific adaptive methods have been proposed, and we could not possibly discuss them all here. However, adaptive methods can be grouped roughly into three major categories. This chapter discusses all three categories in turn. Section 5.2 discusses what are commonly referred to as up/down methods. The basic idea behind up/down methods is straightforward. If the observer responds incorrectly to a trial, the stimulus intensity is increased on the next trial, whereas if the observer responds correctly on a trial (or a short series of consecutive trials), stimulus intensity is decreased on the next trial. In such a procedure, the stimulus level will tend towards a specific proportion correct and oscillate around it once it is reached. While up/down methods work well if one is interested only in the value of the psychometric function's (PFs) threshold, they provide little information regarding the slope of the PF.

Section 5.3 discusses adaptive methods which perform a "running fit" on the data. That is, after every response a PF is fit to the responses of all preceding trials. The stimulus intensity to be used on the next trial is that which corresponds to the best estimate of the PFs threshold, based on all previous trials. As was the case with up/down methods, running fit methods also provide information only about thresholds, not slopes of the PF.

In Section 5.4 we will discuss the "psi method." The psi method combines ideas from several adaptive methods proposed earlier. The psi method selects stimulus intensities on every trial that maximize the efficiency with which not only the threshold, but also the slope of the psychometric function is estimated. The psi method is arguably the most sophisticated of the adaptive methods in use today.

5.2 UP/DOWN METHODS

5.2.1 Up/Down Method

The up/down method was developed initially by Dixon and Mood (1948). We will explain the logic using the same example that Dixon and Mood used, which is that of determining the sensitivity of explosive mixtures to shock. Apparently, it was common practice to do this by dropping weights from different heights on specimens of explosive mixtures, and noting whether an explosion resulted. The idea is that there will be a critical height which, if exceeded, will result in a mixture exploding whereas below this height it will not explode. Let us say we drop a weight from a height of 20 feet and no explosion occurs. We now know that the critical height is greater than 20 feet. We could investigate further by increasing the height in steps of, say, one foot. We drop the weight from 21 feet . . . nothing, 22 feet . . . still nothing, 23 feet . . . Kaboom! We now know that the critical height has a value between 22 and 23 feet.

In reality, things are a little bit more complicated, of course, in that no two explosive mixtures are identical, and no two drops and consequent impacts of a weight are identical either. We are no experts on explosive mixtures, but we imagine that other factors also play a role. Thus, it would be more appropriate to say that for every drop-height there is some probability that it will cause an explosive mixture to explode; the greater the height, the higher the probability that the mixture will explode. We might define the critical height as that height at which a mixture has a probability of, say, 50% of exploding. You will have realized the similarity between the problem of determining such an "explosion threshold" and that of determining, say, a detection threshold in the context of a sensory experiment.

Keeping the above in mind, the most we can conclude for certain from the above experiment is that the probability that a mixture will explode at 23 feet has a value greater than 0, and the probability that it will explode at 22 feet has a value less than 1. In order to get a better idea of what the value of the "explosion threshold" is, we should get more data. From what we have done so far, it seems reasonable to assume that the explosion threshold has a value somewhere around 22 or 23 feet. It would be silly to start dropping weights from a height of 150 feet or 1 foot at this point. The former is almost certain to result in an explosion, the latter is almost certain not to result in an explosion.

Dixon and Mood (1948) suggest a very simple rule to decide which height we should drop a weight from on any trial. The rule simply states that if an explosion occurs on a trial, we should decrease the drop height on the next trial. If an explosion does not occur on a trial, we should increase the height on the next trial. In other words, our decision regarding the height to use on any trial is determined by what happened on the previous trial, and for this reason this method is termed an adaptive method. Figure 5.1 shows the results of a simulated experiment which continues the series started above following the simple up/down rule. Any trials that resulted in an explosion are indicated by the star-shaped, filled symbols, trials that did not are indicated by the open circles. The height corresponding to the 50% threshold that was used in the simulations is indicated in the figure by the broken line. We will discuss how to derive an explosion threshold estimate from these data later (Section 5.2.5), but for now we note that almost all trials (16 of 17 trials, or 94%) where the drop height was 23 feet resulted in an explosion, and that only 7 of

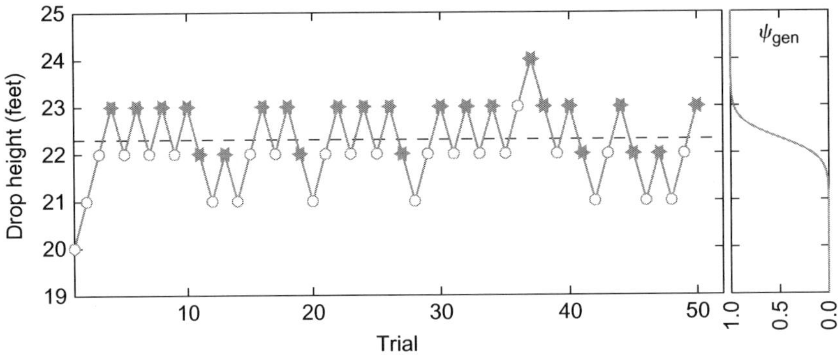

FIGURE 5.1 Simulated run of Dixon and Mood's (1948) up/down method. The figure follows the example described in the text. Weights are dropped from various heights (ordinate) on explosive mixtures. If an explosion occurs (filled symbols) the drop height is reduced by 1 foot on the next trial, if no explosion occurs (open circular symbols), the drop height is increased by 1 foot. The targeted height (22.3 feet) is indicated by the broken line. Responses were generated by a Logistic function with $\alpha = 22.3$, $\beta = 5$, $\gamma = 0$, and $\lambda = 0$. The generating function (ψ_{gen}) is shown on the right.

the 23 trials (30%) where the drop height was 22 feet resulted in an explosion. Thus, it seems reasonable to assume that the explosion threshold has a value somewhere between 22 and 23 feet.

Dixon and Mood's up/down method targets the point on the psychometric function at which either of two responses is equally likely to occur. This makes it particularly useful when one is interested in determining the point of subjective equality in an appearance-based task. For example, in the Muller–Lyer illusion (see Chapter 2) one could use the up/down method to find the ratio of line lengths at which the observer is equally likely to respond with either line when asked to indicate which of the lines appears to be longer.

5.2.2 Transformed Up/Down Method

As mentioned, Dixon and Mood's up/down method targets the stimulus intensity at which either of two possible responses is equally likely to occur. In many experimental situations this will be no good. In a 2AFC task, for example, 50% correct corresponds to chance performance. In such situations, we could use Wetherill and Levitt's (1965) "transformed" up/down method. In the transformed up/down method the decision to decrease stimulus intensity is based on a few preceding trials, rather than the very last single trial. For example, we could adopt a rule that increases stimulus intensity after every incorrect response as before, but decreases stimulus intensity only after two consecutive correct responses have been observed since the last change in stimulus intensity. Such a rule is commonly referred to as a 1 up/2 down rule. Wetherill and Levitt make the argument that the 1 up/2 down rule targets 70.71% correct. Another commonly used rule is similar to the 1 up/2 down except that stimulus intensity is decreased only after three consecutive responses have been observed since the last change in stimulus intensity. This 1 up/3 down rule targets 79.37% correct. A simulated experimental run using the 1 up/2 down rule is shown in Figure 5.2a. Correct responses are indicated by the filled symbols, incorrect responses are shown by the open symbols. Note that the 1 up/2 down rule came into effect only after the first incorrect response was observed. Before this point, a 1 up/1 down rule was employed. In the run shown, the simulated observer responded correctly on all of the first five trials. This was because the run started out at a stimulus intensity which was well above the targeted threshold, and possibly a bit of luck. Either way, had we adopted the 1 up/2 down rule from the start, it would have likely taken many more trials to reach stimulus intensities around threshold levels. The strategy to adopt a 1 up/1 down rule until a first reversal of direction is needed was suggested by Wetherill and Levitt in order to avoid presenting many trials at intensities which are far above threshold at the start of the run.

5.2.3 Weighted Up/Down Method

Another possibility is to adopt a "weighted" up/down method (Kaernbach, 1991) in which a 1 up/1 down rule is used, but the steps up are not equal in size to the steps down. Kaernbach argues that the rule targets a probability correct equal to:

$$\psi_{\text{target}} = \frac{\Delta^+}{\Delta^+ + \Delta^-} \tag{5.1a}$$

where Δ^+ and Δ^- are the sizes of the steps up and steps down, respectively and ψ_{target} is the targeted proportion correct. A little algebra reveals:

$$\frac{\Delta -}{\Delta^+} = \frac{1 - \psi_{\text{target}}}{\psi_{\text{target}}} \tag{5.1b}$$

Let's say you wish to target 75% correct performance. Using a value of 0.75 for ψ_{target} in Equation 5.1b gives:

$$\frac{\Delta^-}{\Delta^+} = \frac{1 - 0.75}{0.75} = \frac{1}{3}$$

Figure 5.2b shows a simulated run of 50 trials using a 1 up/1 down rule and a ratio of stepsizes equal to 1/3.

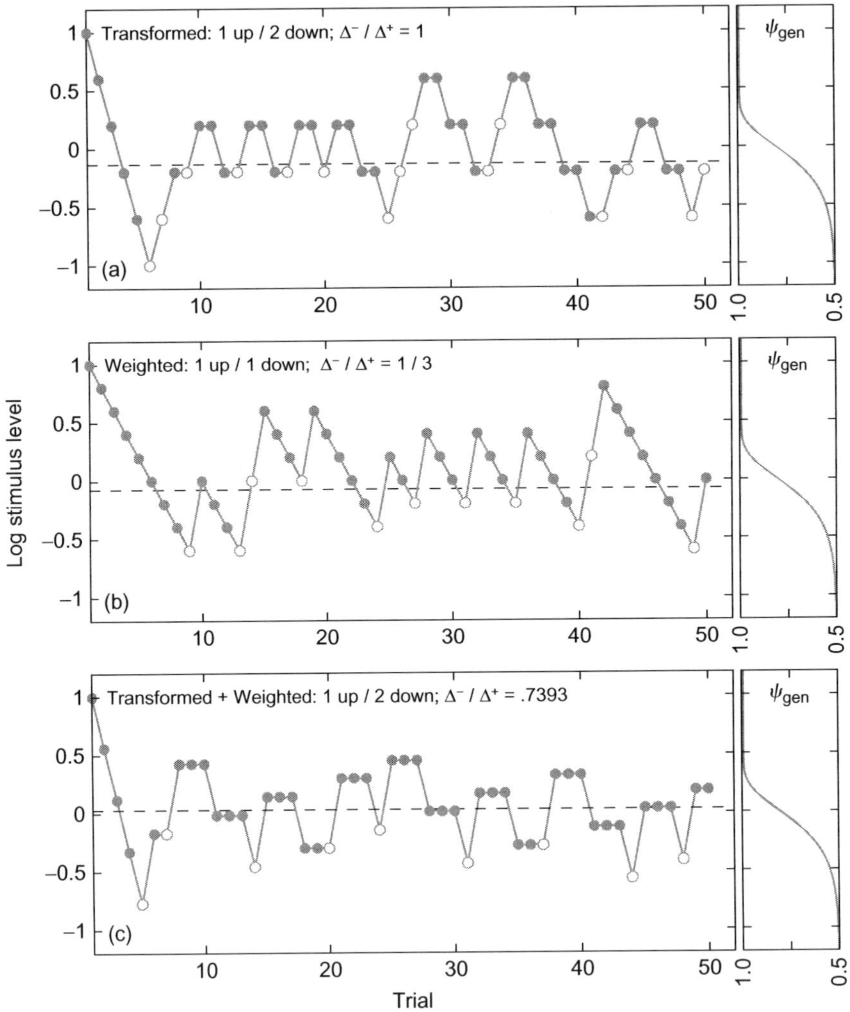

FIGURE 5.2 Examples of simulated staircases following a transformed up/down rule (a); a weighted up/down rule (b); and a transformed and weighted up/down rule (c). Correct responses are indicated by the filled symbols, incorrect responses are indicated by open symbols. Stimulus levels corresponding to the targeted percent correct values are indicated by the broken lines (note that the different procedures target different performance levels). In all example runs the responses were generated by a Gumbel function with $\alpha = 0$, $\beta = 2$, $\gamma = 0.5$, and $\lambda = 0.01$. The generating PF (ψ_{gen}) is shown to the right of each graph. Δ^+: size of step up; Δ^-: size of step down (see Section 5.2.3).

5.2.4 Transformed and Weighted Up/Down Method

In the "transformed and weighted up/down method" (García-Pérez, 1998), steps up and steps down are, as in the weighted method, not of equal size. Also, stimulus intensity is decreased only after a set number of consecutive incorrect responses, as in the transformed up/down method. The proportion correct targeted by the transformed and weighted up/down method is given as:

$$\psi_{\text{target}} = \left(\frac{\Delta^+}{\Delta^+ + \Delta^-}\right)^{\frac{1}{D}} \tag{5.2}$$

where Δ^+, Δ^- and ψ_{target} are as before, and D is the number of consecutive correct responses after which a step down is to be made. Dixon and Mood's up/down method, the weighted up/down method and the transformed up/down method are, of course, also covered by Equation 5.2, as they are all particular cases of the transformed and weighted up/down method. Figure 5.2c shows an example run of 50 trials using a transformed and weighted up/down method.

5.2.5 Termination Criteria and the Threshold Estimate

Several methods are in use to estimate a threshold after a run of trials has completed. Most commonly, researchers will terminate a run after a specific number of reversals of direction have occurred (García-Pérez, 1998). The threshold estimate is consequently calculated as the average stimulus intensity across the last few trials on which a reversal occurred in the run. For example, a run may be terminated after ten reversals have taken place, and the threshold estimate is calculated as the average stimulus intensity across the last eight trials on which a reversal occurred. Less frequently, the run is terminated after a specified number of trials have occurred and the threshold is calculated as the average stimulus intensity across the last so many trials.

Yet another strategy is to adopt what Hall (1981) has termed a "hybrid adaptive procedure" in which an up/down method is used to select stimulus intensities, after which a threshold estimate is derived by fitting a PF to all the data collected using, for example, a maximum likelihood criterion (see Chapter 4). This strategy has the obvious disadvantage that fitting the data with a PF requires us to assume a shape of the PF, as well as values of some of its parameters (such as the guess rate, lapse rate, and perhaps the slope). The up/down methods themselves are non-parametric; they do not assume a particular shape of the underlying PF, other than that it is a monotonic function of stimulus intensity.

After we have determined the thresholds for the individual runs we face another decision. Typically, we would want to use more than one run, and thus we end up with several threshold estimates which should be combined into a single estimate. The obvious, and indeed most common, solution is to average the individual threshold estimates. The standard deviation of threshold estimates may serve the function of standard error of estimate. The hybrid adaptive procedure allows us to combine trials across the different runs before we fit them with a single PF. When we do use the hybrid procedure, we should keep in mind that by the nature of the up/down procedure trials will be concentrated around a single point on the PF. Such data do not lend themselves well to the estimation of the slope of the PF. Combining all data and fitting them with a single PF will also allow us to determine the reliability of our estimate by performing a bootstrap analysis or by using the standard deviation of the parameter's posterior distribution (see Chapter 4).

5.2.6 Up/Down Methods in Palamedes

The core function associated with the up/down methods in Palamedes is **PAL_AMUD_updateUD**. The function **PAL_AMUD_updateUD** is called after every trial with two arguments: a structure which we call **UD** (although you may give it a different name); and a scalar which indicates whether the observer gave a correct (1) or an incorrect (0) response on the trial. The structure **UD** stores things such as the stimulus intensity used on each trial, the response of the observer, etc., and this information is updated after every trial when the function **PAL_AMUD_updateUD** is called. The **UD** structure also stores such things as the up/down rule to be used, stepsizes to be used, the stimulus value to be used on the first trial, etc.

Before trials can begin, the function **PAL_AMUD_setupUD** must be called in order to create the **UD** structure and initialize some of its entries. Let's first create the **UD** structure using the default values and inspect it:

```
>>UD = PAL_AMUD_setupUD;
>>UD
UD =

               up:  1
             down:  3
       stepSizeUp:  0.0100
     stepSizeDown:  0.0100
    stopCriterion:  'reversals'
         stopRule:  32
       startValue:  0
             xMax:  []
             xMin:  []
         truncate:  'yes'
         response:  []
             stop:  0
```

```
             u:  0
             d:  0
     direction:  []
      reversal:  0
      xCurrent:  0
             x:  0
    xStaircase:  []
```

The **up** and **down** fields indicate which up/down rule should be used. By default the values are set to 1 and 3, respectively, such that a 1 up/3 down rule will be used. Of course, the default values can be changed to suit your needs. You can pass optional arguments to **PAL_AMUD_setupUD** to make such changes. These arguments come in pairs. The first argument of each pair indicates which option should be changed, and the second argument of the pair indicates the new value. For example, let's say you would like to use a 1 up/2 down rule instead of the default 1 up/3 down rule. You would change the **UD** field **down** to 2 by giving the following command:

```
>>UD = PAL_AMUD_setupUD(UD, 'down', 2);
```

In this call we also passed the existing **UD** structure as the first argument. When an existing **UD** structure is passed to **PAL_AMUD_setupUD** as the first argument, the existing structure **UD** will be updated according to the additional arguments. It is also possible to create an entirely new structure **UD** using optional arguments to override the default values. For example, the call:

```
>>UD = PAL_AMUD_setupUD('down', 2);
```

creates a new structure **UD**, but with the field **down** set to 2 instead of the default value 3. It is important to realize that when the latter syntax is used, a new structure is created and any previous changes made to **UD** will be undone. Before we worry about the other options, let us demonstrate how to use the function **PAL_AMUD_updateUD** to update the stimulus level on each trial according to the chosen up/down rule. As we go along, we will explain some of the other options.

Imagine we are measuring a contrast threshold for an observer and we would like to use a 1 up/3 down rule. We vary contrast amplitudes on a logarithmic scale and wish to use stepsizes equal to 0.05 log units for steps up as well as steps down. We first create a new **UD** structure with the above options.

```
>>UD = PAL_AMUD_setupUD('up', 1, 'down', 3, 'stepsizeup', ...
0.05, 'stepsizedown', 0.05);
```

We also wish the procedure to terminate after, say, 50 trials have occurred. We can change the relevant settings in the existing **UD** structure by calling the function again and passing it the existing structure as the first argument followed by the other settings we wish to change. We set the value of **stopCriterion** to the string **'trials'** to indicate that we wish to terminate the run after a set number of trials (the default setting is to terminate after a set number of reversals of direction have occurred). Because we wish the run to terminate after 50 trials, we set the value of **stopRule** to 50.

```
>>UD = PAL_AMUD_setupUD(UD, 'stopcriterion', 'trials', ...
'stoprule', 50);
```

We should also indicate what contrast amplitude should be used on the first trial (or accept the default value of 0):

```
>>UD = PAL_AMUD_setupUD(UD, 'startvalue', 0.3);
```

Note that all changes to the default entries could have also been made in a single call, and also that the case of the string arguments will be ignored.

We are now ready to present our first stimulus to our observer. The contrast amplitude that we should use on any trial is given in **UD**'s field **xCurrent**. Currently, the value of **UD.xCurrent** is 0.3. This is, after all, the value that we submitted as **'startvalue'**. After we present our stimulus at the amplitude given in **UD.xCurrent** we collect a response from the observer. We create a variable **response** and assign it the value 1 in case the response was correct, or the value 0 in case the response was incorrect.

```
>>response = 1;
```

Now we call the function **PAL_AMUD_updateUD**, passing it the structure **UD** and the value of **response**:

```
>>UD = PAL_AMUD_updateUD(UD, response);
```

PAL_AMUD_updateUD makes the appropriate changes to **UD** and returns the updated version. You will have noted that the above call assigns the returned structure to **UD**. In effect, the updated **UD** will replace the old **UD**. The value of **UD.xCurrent** has now been updated according to the staircase rules, and should be used as the contrast amplitude to be used on the next trial.

```
>>UD.xCurrent

ans =
0.2500
```

The process then repeats itself: we present the next trial at the new value of **UD.xCurrent**, collect a response from the observer, and call **PAL_AMUD_updateUD** again. When the criterion number of trials has been reached, the **stop** field of the **UD** structure will be set to 1, and this will be our signal to exit the testing procedure. The program that controls your experiment would contain a trial loop such as this:

```
while ~UD.stop
%Present trial here at stimulus intensity UD.xCurrent
%and collect response (1: correct [more generally: too high],
%0: incorrect)
UD = PAL_AMUD_updateUD(UD, response); %update UD structure
end
```

The **UD** structure maintains a record of stimulus intensities and responses for all trials. The stimulus amplitudes of all trials are stored in **UD.x**, and the responses are stored in **UD.response**. The file **PAL_AMUD_Demo** demonstrates how to use the functions in the Palamedes toolbox to implement the 1 up/3 down staircase discussed here. It simulates responses of a hypothetical observer who acts according to a Gumbel function. The program will produce a plot such as those in Figure 5.2. A correct response to a trial is indicated by a filled circle, an incorrect response by an open circle. Note again that at the start of the series, the procedure decreases the stimulus intensity after every correct response. It is only after the first reversal of direction occurs that the 1 up/3 down rule goes into effect.

The field **UD.reversal** contains a 0 for each trial on which a reversal did not occur and the count of the reversal for trials on which a reversal did occur. For example, in the run shown in Figure 5.2a the first reversal took place due to an incorrect response on trial 6, the second reversal took place following the second of two consecutive correct responses on trial 11. Thus, the field **UD.reversal** will start: **[0 0 0 0 0 1 0 0 0 0 2 ....]**.

The function **PAL_AMUD_analyzeUD** will calculate the mean of either a specified number of reversals, or a specified number of trials. By default, it will calculate the average of all but the first two reversal points. Usage of the function is as follows:

```
>>Mean = PAL_AMUD_analyzeUD(UD);
```

where **UD** is the result of a run of an up/down adaptive procedure. You may override the default and have the mean calculated across a specific number of reversals. For example, in order to calculate the mean across the last five reversals use:

```
>>Mean = PAL_AMUD_analyzeUD(UD, 'reversals', 5);
```

You may also calculate the mean across the last so many trials. For example, in order to calculate the mean across the last ten trials use:

```
>>Mean = PAL_AMUD_analyzeUD(UD, 'trials', 10);
```

Keep in mind that all data are stored in the **UD** structure and you may use them as you please. For example, you could use **PAL_PFML_Fit** (Chapter 4) to fit a PF to the data using a maximum likelihood criterion. For example, to fit a threshold value to the data assuming the shape of a Logistic function, a slope of 2, a guess rate of 0.5 and a lapse rate of 0.01 use:

```
>>params = PAL_PFML_Fit(UD.x, UD.response, ones(1, ...
length(UD.x)), [0 2 .5 .01], [1 0 0 0], @PAL_Logistic);
```

5.2.7 Some Practical Tips

Using a large number of simulated up/down staircases, García-Pérez (1998) has investigated the behavior of up/down staircases systematically. Somewhat surprisingly perhaps, the staircases converged reliably on the proportion correct given by Equation 5.2 only when specific ratios of the up and down stepsizes (Δ^-/Δ^+) were used. These ratios are listed in Table 5.1. At other ratios, the proportion correct on which the staircases converged depended greatly on the ratio of the stepsize to the spread of the psychometric function (Section 4.3.2.7). For certain stepsize ratios and up/down rule combinations the staircases converged on proportions correct nowhere near those given by Equation 5.2. Note that the run shown in Figure 5.2c uses the suggested stepsize ratio for the 1 up/3 down rule employed.

Large stepsizes should be used, with steps up having a value between $\sigma/2$ and σ, where σ is the spread of the underlying PF using $\delta = 0.01$ (Section 4.3.2.7). Of course, the spread of the underlying PF will not be known, but we can generate a rough estimate based on intuition or previous research. Large stepsizes produce reversals more quickly and allow for a faster return to stimulus intensities near the targeted threshold after, for example, a series of lucky responses. The use of large stepsizes also ensures that near-threshold levels are reached early in the run. As a result, we may determine the threshold by averaging stimulus intensities across all but the first few reversals.

TABLE 5.1 Ratios of down stepsize and up stepsize Δ^-/Δ^+ that will reliably converge on the targeted ψ values given by Equation 5.2. These values are suggested by García-Pérez (1998) and are based on a large number of simulated runs

Rule	Δ^-/Δ^+	Targeted ψ (%)
1 up/1 down	0.2845	77.85
1 up/2 down	0.5488	80.35
1 up/3 down	0.7393	83.15
1 up/4 down	0.8415	85.84

Stepsizes should be defined in whatever metric appears appropriate. When we started our discussion of up/down methods we used the example of explosive mixtures. We defined the stepsize in terms of drop height (in feet). Consequently, the procedure used steps that were equal in terms of drop height in feet. Perhaps it would have made more sense to use stepsizes that correspond to equal changes in the speed with which the weights hit the explosive mixture. If so, we should simply define our stepsizes in terms of speed at impact. In the context of psychophysical measurements, stepsizes should be defined in physical units that would correspond to linear units in the internal representation of the stimulus dimension. Our choice should thus be guided by what we believe the transducer function to be (see Section 4.2.2.3). Our choice will ordinarily be between defining stepsizes on a linear scale or on a logarithmic scale. If we wish steps to be constant on a logarithmic scale, we should define our stimulus intensities and stepsizes as such. The final threshold should be calculated as the arithmetic mean calculated across the reversal values in whatever scale our stepsizes were defined in. For example, if we defined our stimulus intensities and stepsizes on a logarithmic scale, we should calculate the arithmetic mean of the reversal values in logarithmic terms, which is equivalent to the geometric mean of the values on a linear scale.

Note that you may set a minimum or maximum stimulus value in the **UD** structure (Textbox 5.1). You should avoid this if you can, but sometimes you have no other choice. For example, if stimulus intensity is defined as Michelson contrast on a linear scale, and you do not set the minimum stimulus value to 0, the procedure might assign a negative value to **UD.xCurrent**. Of course, we cannot present stimuli at negative contrast and setting the minimum stimulus value to 0 will avoid assigning a negative value to **UD.xCurrent**.

An issue is raised when possible stimulus values are constrained to a range of values. Suppose you are measuring a contrast threshold. Your stimulus intensities and stepsizes are defined as Michelson contrast on a linear scale. You use a 1 up/1 down with $\Delta^+ = 0.1$ and $\Delta^- = 0.05$. Having defined your stimulus intensity in terms of Michelson contrast you set the minimum stimulus intensity to 0. As it happens your observer has had a few consecutive lucky responses and the stimulus intensity on trial t (let's call it x_t) equals 0. In case your observer responds correctly on trial t, the up/down rule states that stimulus intensity on trial $t + 1$ should be $x_{t+1} = x_t - \Delta^- = 0 - 0.05 = -0.05$. However, having defined the minimum stimulus intensity as 0, the value of **UD.xCurrent** will actually be set to 0.

Imagine the observer now produces an incorrect response on trial $t + 1$. Should we make our step up relative to what the intensity should have been on trial t + 1 ($x_{t+2} = -0.05 + \Delta^+ = 0.05$) or relative to what it actually was ($x_{t+2} = 0 + \Delta^+ = 0.1$)? Somewhat counterintuitively, perhaps, the former strategy has been shown to produce better results (García-Pérez, 1998). You can indicate whether you wish the up/down rule to be applied to stimulus intensities as truncated by the minimum and

Textbox 5.1. Options for the Palamedes up/down routines that may be changed using the function **PAL_AMUD_setupUD** . Default values are those shown in curly brackets {}.

Up positive integer scalar {**1**}

Number of consecutive incorrect responses after which stimulus intensity should be increased.

Down positive integer scalar {**3**}

Number of consecutive correct responses after which stimulus intensity should be decreased.

stepSizeUp positive scalar {**0.01**}

Size of step up

stepSizeDown positive scalar {**0.01**}

Size of step down

stopCriterion **'trials'** | {**'reversals'**}

When set to **'trials'**, staircase will terminate after the number of trials set in **stopRule**. When set to **'reversals'**, staircase will terminate after the number of reversals set in **stopRule**.

stopRule see **stopCriterion** {**32**}

startValue scalar {**0**}

Stimulus intensity to be used on first trial.

xMax scalar { **[]** }

Maximum stimulus intensity to be assigned to **UD.xCurrent**. In case value is set to an empty array (**[]**) no maximum is applied.

xMin scalar { **[]** }

Minimum stimulus intensity to be assigned to **UD.xCurrent**. In case value is set to an empty array (**[]**) no minimum is applied.

truncate **{'yes'}** | **'no'**

When set to **'yes'**, up/down rule will be applied to stimulus intensities as limited by **xMax** and **xMin**. When set to **'no'**, up/down rule will be applied to stimulus intensities untruncated by **xMax** and **xMin** (but stimulus intensities assigned to **UD.xCurrent** will be truncated by **xMax** and **xMin**).

maximum values by setting the value of **'truncate'** to **'yes'**. In case you wish to allow stimulus values to go beyond the minimum and maximum as far as application of the up/down rule is concerned, you should set **'truncate'** to **'no'**. Beside the field **UD.x**, which keeps track of stimulus intensities that are actually used, the **UD** structure has another field (**UD.xStaircase**) which contains for each trial the stimulus intensity to which the up/down rule is applied. In case **'truncate'** is set to **'yes'**, the two fields will have identical entries.

By the very nature of the up/down procedures, strong trial-to-trial dependencies within a single staircase exist. Observers are very good at discovering rules such as: "A series of consecutive lucky guesses is followed by one or more trials on which I also feel like I am guessing. This continues until I give a few incorrect responses." Some evidence even suggests that humans can discover such simple contingency rules implicitly and begin to act accordingly before the rules are consciously known (Bechara et al., 1997). In order to avoid trial-to-trial dependencies and observer strategies which are based on trial-to-trial contingencies, it is a good idea to alternate trials randomly between a few interleaved up/down staircases.

5.3 "RUNNING FIT" METHODS: THE BEST PEST AND QUEST

The methods that we will describe here perform a running fit of the results. The idea was first proposed by Hall (1968) at a meeting of the Acoustical Society of America. After every trial a psychometric function is fit to all the data collected so far. The fitted PF then serves to select a stimulus intensity for the upcoming trial. After each trial, the fit is updated based on the new response and the process repeats itself.

5.3.1 The Best PEST

The first running fit method to be proposed in detail was the "best PEST" (Pentland, 1980). The best PEST assumes a specific form of the psychometric function and estimates only the threshold parameter of the psychometric function. Values for the other parameters (slope, guess rate, and lapse rate) need to be assumed. After each trial, the likelihood function (Chapter 4) is calculated based on the responses to all previous trials. The likelihood function is defined across a range of possible threshold values believed to include the observer's threshold value. After each trial a value for the threshold parameter is estimated using a maximum likelihood criterion. The stimulus intensity to be used on the next trial corresponds to the threshold estimate determined from all previous trials.

A simulated example run of the best PEST in a 2AFC procedure is shown in Figure 5.3a. As can be seen from the figure, a run controlled by the best PEST

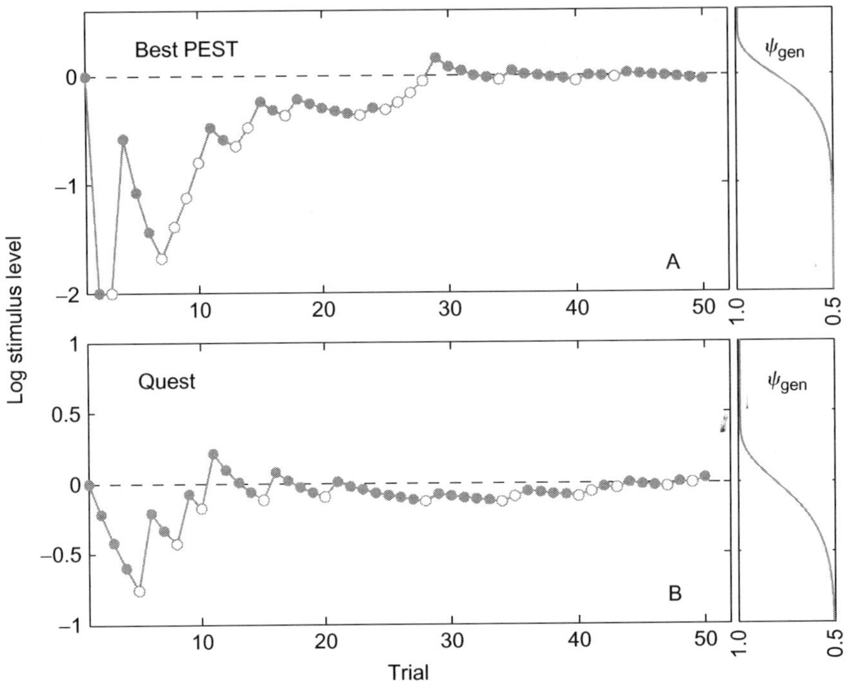

FIGURE 5.3 Examples of a simulated best PEST staircase (a) and a Quest staircase (b). Correct responses are indicated by the filled symbols, incorrect responses are indicated by open symbols. Stimulus levels corresponding to the generating PFs threshold are indicated by the broken lines. In both example runs the responses were generated by a Gumbel function with $\alpha = 0$, $\beta = 2$, $\gamma = 0.5$, and $\lambda = 0.01$. The generating PF (ψ_{gen}) is shown to the right of each graph.

has some similarities to runs controlled by the up/down procedures discussed in Section 5.2. Foremost, the best PEST decreases stimulus intensity after a correct response and increases stimulus intensity after an incorrect response. Keep in mind, though, that in the case of the best PEST this is an emergent property, it is not a rule that is explicitly incorporated in the procedure. Unlike in the up/down procedures of Section 5.2, however, stepsizes in the best PEST are not of fixed size. Generally, stepsizes tend to decrease as the run proceeds. This makes sense when one considers that the relative contribution of each additional trial to the overall fit becomes smaller and smaller, given that the fit is based on *all* of the preceding trials.

Note that the first stepsize is exceptionally large. As a matter of fact, the size of the first step is bound only by the interval of stimulus values defined by the experimenter. The maximum likelihood estimate of the threshold after a single trial will always be positive infinity (if the response was incorrect) or negative infinity (if the response was correct), and thus the stimulus amplitude on the second trial will always correspond to the highest or lowest value of the interval across which the likelihood function is considered. In the example shown in the figure, the first trial resulted in a correct response and the second trial is presented at the lowest stimulus intensity in the interval across which the likelihood function is calculated. The response on the second trial is also correct and, judging by the generating PF on the right of the figure, this should be considered mostly a result of luck. As a result, the third trial is also presented at the same extremely low stimulus intensity. The response to the third trial was incorrect, and as a result the fourth trial is presented at a much higher stimulus intensity, although still at a stimulus intensity where we would expect near-chance-level performance. The following few trials once again appear to be the result of some luck. As a result, it takes a while for the best PEST to reach stimulus intensities near threshold in this particular run. It should be pointed out that this particular run was selected because it displayed this behavior. However, although it may not be typical, this behavior is certainly not uncommon using the best PEST.

5.3.2 Quest

Quest (Watson & Pelli, 1983) is essentially a Bayesian version of the best PEST. To refresh your memory, Bayes' Theorem (Chapter 4) can be used to combine the results of an experiment (in the form of the likelihood function) with pre-existing knowledge or beliefs regarding the value of the threshold parameter (in the form of a prior probability distribution) to derive the posterior probability distribution across possible values of the threshold parameter. One may think of this procedure as the new data merely serving to adjust our pre-existing knowledge or beliefs regarding the value of the threshold parameter. A best-fitting estimate of the threshold parameter is then derived from the posterior distribution using the method outlined in Chapter 4. Thus, before a Quest run can start, the researcher needs to postulate a prior probability distribution which reflects the researcher's belief about the value of the threshold. In Figure 5.3b, we show an example Quest run in which the prior distribution was a Gaussian distribution with mean 0 and standard deviation 1. The simulated observer in the Quest run of Figure 5.3 had identical properties to the simulated observer in the best PEST run in the figure. As can be seen by comparing the best PEST and Quest run, the prior has the effect of curbing the excessive stepsizes that were observed at the beginning of the best PEST run. The prior could thus be considered to act as somewhat of a guide to the selection of stimulus intensities. This is especially true at the onset of the run when stimulus selection is primarily determined by the prior. As data collection proceeds, the prior will start to play a smaller and smaller role relative to the contribution of the data collected.

In the example run of Quest shown in Figure 5.3, the mode of the posterior distribution was used as the threshold estimate, as proposed by Watson and Pelli in the original Quest procedure. King-Smith et al. (1994) show that using the mean of the posterior distribution, rather than its mode, leads to more efficiently obtained parameter estimates which are also less biased. Alcalá-Quintana and García-Pérez (2004) further recommend the use of a uniform prior (i.e., one that favors no particular threshold value over another). Remember that when a uniform prior is used, the posterior distribution will be proportional to the likelihood function (see Chapter 4). Thus, when the prior is uniform and we use the mode of the posterior distribution as our estimate of the threshold, Quest is equivalent to the best PEST.

5.3.3 Termination Criteria and Threshold Estimate

Most commonly, a session is terminated after a specific number of trials. Alternatively one can terminate a session after a specific number of reversals have occurred. The threshold estimate is, of course, updated every trial and the final threshold estimate is simply the estimate which was derived following the final response. In case we have used a non-uniform prior, we may opt to ignore it in our final threshold estimate. Remember from Chapter 4 that the posterior distribution is proportional to the product of the prior distribution and the likelihood function. In other words, if we divide the prior out of the posterior distribution the result is proportional to our likelihood function. Since choosing a prior is a bit of a subjective exercise, some researchers opt to derive the final threshold estimate from the (recovered) likelihood function. As with the up/down methods we may also use a hybrid approach in which we use a running fit method to guide stimulus selection, but we derive our final threshold estimate and its standard error after we combine trials across several sessions.

5.3.4 Running Fit Methods in Palamedes

The routines in Palamedes that manage a running fit adaptive method are **PAL_AMRF_setupRF** and **PAL_AMRF_updateRF**. The general usage of these functions is analogous to the functions **PAL_AMUD_setupUD** and **PAL_AMUD_updateUD** described in Section 5.2.6. We first create a structure **RF** using **PAL_AMRF_setupRF**.

```
>>RF = PAL_AMRF_setupRF;
```

RF is a structure which is similar to the structure **UD** in Section 5.2.6.

```
>>RF
RF =

        priorAlphaRange:   [1×401 double]
                  prior:   [1×401 double]
                    pdf:   [1×401 double]
                   mode:   0
```

```
              mean:  1.9082e-017
                sd:  1.1576
  modeUniformPrior:  []
  meanUniformPrior:  []
    sdUniformPrior:  []
          response:  []
     stopCriterion:  'trials'
          stopRule:  50
              stop:  0
                PF:  @PAL_Gumbel
              beta:  2
             gamma:  0.5000
            lambda:  0.02
              xMin:  []
              xMax:  []
         direction:  []
          reversal:  0
          meanmode:  'mean'
          xCurrent:  1.9082e-017
                 x:  []
        xStaircase:  1.9082e-017
```

The value of the **mean** field is calculated as the expected value of the prior distribution and differs (very slightly) from zero due to rounding error only. As a result, **xCurrent** and **xStaircase** also differ slightly from zero. Changing the settings to suit your needs is done in a manner similar to changing the values in the **UD** structure in Section 5.2. Let's say we wish to specify the prior to be something different from the uniform prior (which is the default). We must specify the range and resolution of values of possible thresholds to be included in the prior (or accept the default range and resolution: **-2:.01:2)** and define a prior distribution across that range:

```
>>alphas = -3:.01:3;
>>prior = PAL_pdfNormal(alphas, 0, 1);
```

The above call to **PAL_pdfNormal** returns the normal probability densities at the values in **alphas** using a mean equal to 0 and standard devation equal to 1. Next, we update the relevant fields in **RF**:

```
>>RF = PAL_AMRF_setupRF(RF, 'priorAlphaRange', alphas, ...
'prior', prior);
```

The **RF** options and their default values are given in Textbox 5.2.

Textbox 5.2. Options for the Palamedes running fit routines that may be changed using the function **PAL_AMRF_setupRF**. Default values are those shown in curly brackets {}.

priorAlphaRange vector {**[-2:.01:2]**}

Vector containing values of threshold to be considered in fit.

prior vector {uniform across **priorAlphaRange**}

Prior distribution.

beta positive scalar {**2**}

Slope parameter of PF to be fitted.

gamma scalar in range [0–1] {**.5**}

Guess rate to be used in fits.

lambda scalar in range [0–1] {**.02**}

Lapse rate to be used in fits.

PF inline function {**@PAL_Gumbel**}

Form of psychometric function to be used in fit. Refer to Section 4.3.2 for other possible functions.

stopCriterion {**'trials'**} | **'reversals'**

When set to **'trials'**, staircase will terminate after the number of trials set in **stopRule**. When set to **'reversals'**, staircase will terminate after the number of reversals set in **stopRule**.

stopRule positive integer {**50**}

see **stopCriterion**

startValue scalar {**0**}

Stimulus intensity to be used on first trial.

xMin scalar {**[]**}

Minimum stimulus intensity to be assigned to **RF.xCurrent**. If set to empty array, no minimum will be applied.

xMax scalar {**[]**}

Maximum stimulus intensity to be assigned to **RF.xCurrent**. If set to empty array, no maximum will be applied.

meanmode {**'mean'**} | **'mode'**

Indicates whether the mean or the mode of the posterior distribution should be assigned to **RF.xCurrent**.

During the testing session, the function **PAL_AMRF_updateRF** updates the posterior distribution after each trial and keeps a record of the stimulus intensities, responses, etc. In the code that controls our experiment we would have a loop:

```
while ~RF.stop
amplitude = RF.xCurrent; %Note that other value may be used
%Present trial here at stimulus intensity 'amplitude'
%and collect response (1: correct, 0: incorrect)
RF = PAL_AMRF_updateRF(RF, amplitude, response); %update RF
end
```

Note that we also pass the stimulus intensity (**amplitude**) to **PAL_AMRF_updateRF** (we did not do this with the up/down routines). This is because we are entirely free to ignore the value suggested by the procedure and present the stimulus at some other intensity. As such, we need to tell the procedure what stimulus intensity we actually used.

The **RF** structure stores the stimulus intensities that were actually used on each trial in the field **RF.x**, and the corresponding responses in the field **RF.response**. The stimulus intensities that were suggested by the procedure on each trial are stored in the field **RF.xStaircase**. The final estimates of the threshold in the **RF** structure are **RF.mode** (the mode of the posterior distribution), and **RF.mean** (the mean of the posterior distribution). **RF.sd** contains the standard deviation of the posterior distribution. This standard deviation may serve as the standard error of estimate of the threshold if the threshold is estimated by the mean of the posterior distribution. The entries **RF.modeUniformPrior**, **RF.meanUniformPrior** and **RF.sdUniformPrior** are analogous, except that they ignore the prior distribution provided by the researcher and instead use a uniform prior.

5.3.5 Some Practical Tips

The running fit methods described here are "parametric methods." What this means is that they assume that our observer responds according to a specific form of PF with a specific value for its slope, guess rate, and lapse rate. We need to specify all these assumptions, and the method is optimal and accurate only insofar as these assumptions are true. This was not the case for the up/down methods of Section 5.2. There, we do not have to assume anything about the shape of the PF (other than that it is monotonic). The running fit methods appear a bit awkward perhaps, because we pretend to know all about the observer's PF except for the value of the threshold. As it turns out, though, the procedures are relatively robust when inaccurate assumptions regarding the PF are used. Nevertheless, we should use our best efforts to have our assumptions reflect the true state of the world as accurately as possible. We might base our guesses on our experience with similar experimental conditions, or we could perform some pilot experiments first.

The value we use for the slope affects the stepsizes that the running fit methods use. When a value for the slope is used that is much too high, the methods will use very small stepsizes. As a result, the method becomes sluggish, in that it will have a relatively hard time recovering from a series of lucky responses at very low stimulus intensities, for example. We might also consider allowing for some lapses to occur by setting the lapse parameter to a small value, such as 0.02 (which is the default). The risk you run by setting the lapse rate to 0 is that when a lapse does occur at a high stimulus intensity, it will be followed by a series of trials at intensities well above threshold.

It is not necessary to present the stimulus on any trial at the intensity suggested by the running fit method. For example, you might prefer to present a stimulus at an intensity which is a bit higher than the threshold intensity in order to avoid frustration on the part of the observer. Presenting a stimulus at a high intensity every once in a while will remind the observer what to look for and might act as a confidence booster. You might also target a few distinct points along the PF so as to generate data that are suitable to estimate the threshold as well as the slope of the PF (although we recommend you to keep reading and use the psi method if your goal is to estimate the slope of the PF as well as the threshold).

Here, as with the up/down methods, it is a good idea to intertwine a few staircases randomly to avoid trial-to-trial dependencies. It is always possible to combine observations from the different staircases later and fit all simultaneously with a single PF, as in Hall's (1981) hybrid procedure (Section 5.2.5).

The choice of the prior to use deserves some attention. Remember that even if you choose to use a uniform prior, it is in practice not truly uniform as it will be defined across a finite range of values (by your choice of values to include in **priorAlphaRange** in Palamedes). In other words, threshold values within the finite range of values are given equal likelihoods, but values outside of that range are assigned a likelihood of 0. It is important to let your threshold estimates not be affected significantly by your choice of the range of the prior. For example, if you use the mode of the posterior distribution, make sure that the value of the mode is in the range of values included within the prior and not at either boundary of the prior. When you use the mean of the posterior distribution as your threshold estimate, make sure that the posterior distribution is (effectively) contained entirely within the range of values in the prior, at least when your final estimate of the threshold is made. In case the posterior distribution is chopped off abruptly by the limits of the prior, your choice of these limits will have a significant effect on the threshold estimate.

In case observers are tested in multiple sessions in the same experimental conditions, we advise the use of the posterior distribution resulting from the previous session as the prior for a new session. In a sense, the staircase will proceed from session to session as if data were collected in a single session. The MATLAB® file that demonstrates the RF routines (**PAL_AMRF_Demo**) shows how to accomplish this. In effect, at the end of a session the **RF** structure is saved to disc. Before the next session starts, the **RF** structure is loaded from the disc and used in the new session.

Caution should be exercised when there is reason to suspect that sensitivity varies from session to session, for example due to learning or fatigue. In such situations it might be best to start consecutive sessions with identical priors.

5.4 PSI METHOD

The up/down and running fit methods of Sections 5.2 and 5.3 both target a single point on the psychometric function. As such, data collected by these methods do not provide much information regarding the slope of the psychometric function. However, one may also be interested in determining the slope of the psychometric function. For example, when simulations are to be used to determine standard errors (Chapter 4) or the statistical significance of a model comparison (Chapter 8), it is critical to derive an accurate estimate of the slope of the psychometric function. This is because the slope of the PF ultimately describes the noisiness of the results. For example, if we overestimate the value of the slope of the PF and we simulate the experiment using this biased slope, the simulated observer will produce cleaner results compared to our human observer, and we would underestimate the standard error of the threshold estimate. One may also be interested in the value of the PFs slope for its own sake. As mentioned, the slope describes the noisiness of the results and this noisiness is in part due to noise which is internal to the observer. Research questions regarding the effect of, for example, attention or perceptual learning on the noisiness of the perceptual system may be answered by investigating effects on the slope of the PF.

The first adaptive method designed to assess both the threshold and the slope of a PF was Adaptive Probit Estimation (APE; Watt & Andrews, 1981). In APE, estimates of the location and slope parameters of the PF are obtained periodically during testing through probit analysis. The estimates are based on a fixed number of immediately preceding trials and are used to adjust the selection of stimulus levels to be used on subsequent trials. King-Smith and Rose (1997) proposed the modified ZEST method which updates a posterior distribution across a two-dimensional parameter space after each response. Here, we will discuss the psi method (Kontsevich & Tyler, 1999) in some detail. Currently, the psi method is arguably the most efficient of the adaptive methods which target both an estimate of the location and the slope parameter of the PF.

5.4.1 The Psi Method

The psi method (Kontsevich & Tyler, 1999) is a sophisticated method which selects stimulus amplitudes so as to result in efficient estimation of both the threshold and the slope parameter of a psychometric function. In many ways, it is similar to the

Quest procedure. After each response, the psi method updates a posterior distribution, but now the posterior distribution is defined not only across possible threshold values, but also across possible values of the slope parameter. We have discussed such posterior distributions in Chapter 4, and examples are shown in Figures 4.15 and 4.16. As such, the psi method is similar to King-Smith and Rose's (1997) modified ZEST method which also defined the posterior distribution across possible values of the threshold and the slope parameter. In the modified ZEST method, estimates for both the threshold and slope parameters are continuously modified. Stimulus levels are selected to correspond to specific probabilities of a correct response based on the current estimate of the PF. The psi method, however, will select that stimulus intensity for the upcoming trial which minimizes the expected entropy in the posterior distribution after that trial. The use of entropy as the metric in which to define the amount of information gained from a trial was proposed by Pelli (1987), and was used to optimize information gained regarding an observer's membership in a categorical classification by Cobo-Lewis (1997).

The psi method combines some quite complex issues. In order to break it down a bit, consider the following analogy. Imagine you are in a casino and you face a choice between two rather simple games: "Pick a Queen" and "Grab a Spade." In Pick a Queen you draw a card from a standard deck randomly. If you draw a queen, you win the game and receive \$26, but if you draw something other than a queen, you lose the game and pay \$3. In the game Grab a Spade, you pick a random card from a regular deck of cards and if it is a spade you win the game and receive \$20, but if it is not a spade you lose the game and pay \$8. Which game should you pick? One way to decide is to figure the expected monetary gain of each game. In the game Pick a Queen there is a 1/13 chance that you will indeed pick a queen, consequently win the game and gain \$26. However, there is a 12/13 chance that you lose the game and pay \$3. The expected value of your monetary gain (x) in dollars is:

$$E(x) = \frac{1}{13} \times 26 + \frac{12}{13} \times (-3) = -\frac{10}{13} \approx -0.77$$

You can think of the expected gain as your average gain per game if you were to play this game an infinite number of times. Note that your expected gain is negative. You are, after all, in a casino and casinos only offer games for which your expected monetary gain is negative. In a similar fashion, you can figure the expected monetary gain in the game Grab a Spade:

$$E(x) = \frac{1}{4} \times 20 + \frac{3}{4} \times (-8) = -1$$

The game Pick a Queen has a higher expected monetary gain (albeit still negative) so you choose to play Pick a Queen.

The strategy utilized by the psi method to decide which stimulus intensity to use on any trial is very similar. Where we are faced with a choice between two games to play, the psi method is faced with a choice between various stimulus levels to be used on the next trial. And where we select that game which maximizes our expected monetary gain, the psi method selects that stimulus intensity which minimizes the expected "entropy" in the posterior distribution.

The term entropy is used here as it is defined in the context of information theory (i.e., so-called Shannon entropy). Entropy in this context is a measure of uncertainty. A simple example will demonstrate the concept. Imagine a game in which a card is randomly drawn from a standard deck of cards and you are to determine the suit of the card. On any given draw there is, of course, a probability equal to 1/4 that a heart is drawn, 1/4 that a spade is drawn, etc. We can express the degree of uncertainty with regard to the suit of a randomly drawn card by the entropy H:

$$H = -\sum_i p_i \log_2 p_i \tag{5.3}$$

where i enumerates the four possible suits and p_i stands for the probability that the card is of suit i. So, in the above scenario the entropy is:

$$H = -\left(\frac{1}{4}\log_2\left(\frac{1}{4}\right) + \frac{1}{4}\log_2\left(\frac{1}{4}\right) + \frac{1}{4}\log_2\left(\frac{1}{4}\right) + \frac{1}{4}\log_2\left(\frac{1}{4}\right)\right) = 2$$

Since we used the logarithm with base 2 to calculate the entropy, the unit of measurements is "bits." As such, you can think of the entropy as the number of (smart) yes/no questions that stand between the current situation and certainty (i.e., knowing for sure which suit the card is from). Imagine that you get to ask a yes/no question regarding the suit of the card. Let's say you ask the question: "Is the color of the suit red?" Whether the answer is "yes" or "no," it will reduce uncertainty by half. For example, let's say the answer is "yes." We now know that the card is either a heart or a diamond with equal probability. The entropy becomes:

$$H = -\left(\frac{1}{2}\log_2\left(\frac{1}{2}\right) + \frac{1}{2}\log_2\left(\frac{1}{2}\right) + 0\log_2(0) + 0\log_2(0)\right) = 1$$

(Note that we have defined 0 $\log_2(0)$ to equal 0, as $\lim_{p\downarrow 0} p\log(p) = 0$). In words, the answer to the question reduced the uncertainty from 2 bit to 1 bit. We need one more question to attain certainty. For example, in case we ask: "Is the card a heart?" and the answer is "no," we know for certain that the card must be a diamond. Numerically, entropy would indeed be reduced by another bit to equal zero:

$$H = -(1\log_2(1) + 0\log_2(0) + 0\log_2(0) + 0\log_2(0)) = 0$$

reflecting the absence of uncertainty altogether.

In the context of the psi method, entropy measures the uncertainty associated with the values of the threshold and slope of the PF. From the posterior distribution we get the probability associated with each pair of threshold value (a) and slope value (b) contained in the posterior distribution (let's call this $p(a, b)$ here) and calculate the entropy thus:

$$H = -\sum_a \sum_b p(a,b) \log_2 p(a,b) \tag{5.4}$$

By decreasing the entropy in the posterior distribution one would increase the precision of the parameter estimates.

Thus, the psi method considers a range of possible stimulus intensities to use on the next trial (compare: games to play) and for each calculates what the probabilities of a correct response and incorrect response are (compare: probability of win or loss of game). It also considers the entropy (compare: monetary outcome) which would result from both a correct response and an incorrect response. From these, the psi method calculates the expected entropy (compare: expected monetary gain). It then selects that stimulus intensity that is expected to result in the lowest entropy.

Note that stimulus intensity is a continuous variable such that, in theory, the psi method has the choice between an infinite number of values. In practice, the psi method chooses a stimulus intensity from a relatively large number of discrete stimulus intensities in a specified range. The principle is the same compared to choosing one of the two card games to play, however. That is, the expected entropy is calculated for all possible discrete stimulus intensities and the psi method selects that intensity which will lead to the highest expected entropy.

"Wait a minute!" you may have thought, "How does the psi method know what the probability of a correct response is for any given stimulus intensity? Wouldn't the psi method need to know what the PF is in order to do that? Isn't that exactly what we are trying to figure out?" Indeed, the problem that the psi method faces is a bit more daunting than our problem of deciding which game to play. In deciding which game to play, we know what the probability of a win or loss is for each game. The probability of winning Pick a Queen is 1/13, that of winning Grab a Spade is 1/4. The psi method, however, does not know what the probability of a correct or incorrect response is on any given trial. Instead, on each trial it estimates what these probabilities might be, based on the outcomes of all previous trials. It is as if we were to choose between Pick a Queen or Grab a Spade not knowing how many queens or spades are in the deck. However, as we play the games and witness the outcomes of the draws, we start to get an idea as to the make-up of the deck of cards, and we use this to adjust our choice of game.

Specifically, from the posterior distribution and using a Bayesian criterion (Chapter 4), the psi method finds the best-fitting PF to the responses collected on

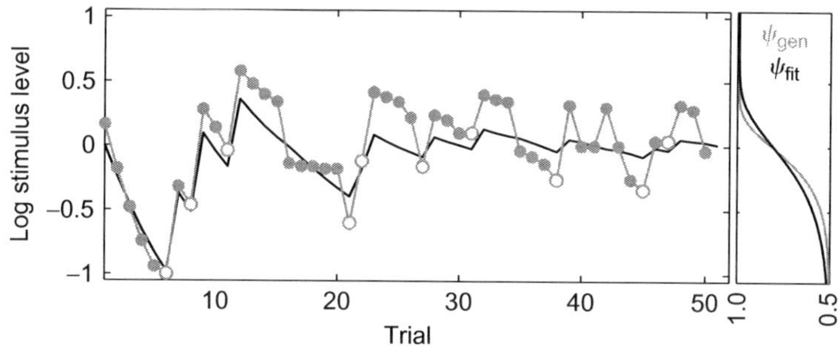

FIGURE 5.4 Example of a simulated psi method staircase. Correct responses are indicated by the filled symbols, incorrect responses are indicated by open symbols. The black line displays the running estimate of the threshold based on the posterior distribution. The responses were generated by a Gumbel function with $\alpha = 0$, $\beta = 2$, $\gamma = 0.5$, and $\lambda = 0.01$. The generating function (ψ_{gen}) is shown on the right in green, the function fitted by the psi method (ψ_{fit}) after completion of the 50 trials is shown on the right in black.

all of the previous trials in the staircase. The probability of a correct response on the next trial at each of the stimulus intensities under consideration is then determined from the best-fitting PF.

An example run in which stimulus intensity was guided by the psi method is shown in Figure 5.4. The same plotting conventions are used here as in Figures 5.2 and 5.3, except that the threshold estimates on each trial (black line) are also shown here. A couple of observations should be made. First, it is apparent that, at the start of the run, the psi method selects stimulus intensities that are at or near the current threshold estimate. This is, of course, the best placement rule to determine the value of the threshold. However, in order to gain information regarding the slope of the PF, measurements need to be made at multiple points along the PF. Indeed, as the run proceeds, the psi method also starts to select stimulus intensities well above and below the running threshold estimate.

5.4.2 Termination Criteria and the Threshold and Slope Estimates

When the psi method is used, it would make no sense to use the number of reversals as a termination criterion. In the up/down and running fit methods, reversals occur mainly when the stimulus amplitude has crossed the targeted threshold. As such, the number of reversals is closely tied to the degree of accuracy with which the threshold can be determined. In the psi method, there is no such close relationship. From Figure 5.4, the pattern with which stimulus amplitudes are selected and reversals occur in the psi method appears much more haphazard compared to the up/down and running fit methods. For example, stimulus amplitude may be increased after a correct response (e.g., after the correct response on trial 41 in Figure 5.4).

Thus, when the psi method is used, a run is terminated after a certain number of trials have occurred.

Most naturally, one would use the posterior distribution created during a run to derive an estimate of the threshold and the slope, and their standard errors. This would be done by the methods described in Chapter 4. However, as with the other adaptive methods, one is free to collect data using the psi method, and consequently use any method to derive one's final parameter estimates. One specific suggestion might be to divide out the prior before we make our final parameter estimates, for the same reasons we discussed in the context of the Quest procedure.

5.4.3 The Psi Method in Palamedes

The functions that implement the psi method in the Palamedes toolbox are used in a fashion very similar to those that implement the up/down methods and the running fit methods. We first set up a structure **PM** by calling **PAL_AMPM_setupPM**:

```
>>PM = PAL_AMPM_setupPM;
```

In case we wish to change some of the options, we can of course do so (for a listing of options see Textbox 5.3). Let's say we wish the testing to terminate after 50 trials, we would like the threshold values to be included in the posterior distribution to be **[-2:.1:2]**, we would like the psi method to consider stimulus intensities from the vector **[-1:.2:1]**, and we wish to accept the default values for the other options. We can change the options in the **PM** structure by calling **PAL_AMPM_setupPM**:

```
>>PM = PAL_AMPM_setupPM('priorAlphaRange', [-2:.1:2], ...
'stimRange', [-1:.2:1], 'numtrials', 50);
```

The program that controls the experiment would contain a loop such as this:

```
while ~PM.stop
%Present trial here at stimulus intensity PM.xCurrent
%and collect response (1: correct, 0: incorrect)
PM = PAL_AMPM_updatePM(PM, response); %update PM structure
end
```

Note that on each trial the stimulus has to be presented at the intensity indicated in the entry in the field **PM.xCurrent**. However, this field will always contain a value that is in the **PM.stimRange** vector which is under our control. Thus, if (for whatever reason) we can (or wish to) present stimuli only at intensities -2, -1, -0.5, 0, 1/3 and π, we need to make that clear to the psi method beforehand. We do that by defining the vector **PM.stimRange** accordingly:

```
PM = PAL_AMPM_setupPM('stimRange', [-2 -1 -.5 0 1/3 pi]);
```

Textbox 5.3. Options for the Palamedes psi method routines that may be changed using the function **PAL_AMPM_setupPM**. Default values are those shown in curly brackets {}.

priorAlphaRange vector {**[-2:.05:2]**}

Vector containing values of threshold to be considered in posterior distribution.

priorBetaRange vector {**[-1:.05:1]**}

Vector containing log transformed values of slope to be considered in posterior distribution.

stimRange vector {**[-1:.1:1]**}

stimulus values to be considered on each trial.

prior matrix {uniform across **priorAlphaRange x prior-BetaRange**}

Prior distribution.

gamma scalar in range [0–1] {0.5}

Guess rate to be used in fits.

lambda scalar in range [0–1] {.02}

Lapse rate to be used in fits.

PF inline function {**@PAL_Gumbel**}

Form of psychometric function to be assumed by psi method. Refer to Section 4.3.2 for other possible functions.

numTrials positive integer {**50**}

Length of run in terms of number of trials.

The psi method will now only suggest values that we can actually present. The Palamedes file **PAL_AMPM_Demo** demonstrates use of the psi method routines. While the demonstration program runs, it will display the posterior distribution after every trial. You'll note that as the session proceeds, and more information is obtained, the posterior distribution will get narrower and narrower.

5.4.4 Some Practical Tips

Many of the practical tips we gave for the running fit methods apply to the psi methods for the same reasons. Here one would also want to allow for lapses to occur by setting the lapse rate to a small non-zero value, such as 0.02. One should also define the prior distribution across ranges of threshold and slope values which are wide enough to accommodate the entire posterior distribution, at least at the time when we derive our final parameter estimates, so as not to let our estimates be determined in large part by the ranges of values we happened to have included in the prior.

The psi method is quite taxing on the RAM memory of your computer. Three arrays of size **length(priorAlphaValues) x length(priorBetaValues) x length(StimRange)** will be created and reside in your RAM memory. Each entry in these arrays will use 8 bytes of RAM. In other words, the amount of RAM memory required to store these matrices alone will be: **3 x 8 x length(priorAlphaValues) x length(priorBetaValues) x length(StimRange)** bytes. Also, each call to **PAL_AMPM_UpdatePM** will involve quite a few calculations. What this means in practical terms is that you need to find an acceptable balance between the resolution of your posterior distribution and possible stimulus values on the one hand, and the time you will allow to perform the necessary computations between trials and the amount of RAM memory you have available on the other.

Exercises

1. By using a program similar to **PAL_AMUD_Demo**, try out what happens when the stepsizes are much smaller or greater than suggested by the text.
2. By using a program similar to **PAL_AMRF_Demo**, try out what happens when the assumed value for the slope differs significantly from the true "generating" slope. Use slopes that are much too high and much too low.
3. By using a program similar to **PAL_AMRF_Demo**, try out what happens when the assumed value for the lapse rate equals 0 but lapses do in fact occur. How does this affect the value for the threshold and slope parameter estimates? What if the assumed value for the lapse rate does not equal 0 (say it equals 0.01) but lapses in fact do not occur?
4. Somebody draws a card from a standard deck. You are to guess what suit it is. Before you have to make a guess, you get to ask one yes/no question. Show that the question "Is the color of the suit red?" results in a lower expected entropy compared to the question "Is the suit hearts?"
5. Repeat question 3 in the context of the psi method.

References

Alcalá-Quintana, R., & García-Pérez, M. A. (2004). The role of parametric assumptions in adaptive Bayesian estimation. *Psychological Methods*, *9*, 250–271.

Bechara, A., Damasio, H., Tranel, D., & Damasio, A. R. (1997). Deciding advantageously before knowing the advantageous strategy. *Science*, *275*, 1293–1295.

Cobo-Lewis, A. B. (1997). An adaptive psychophysical method for subject classification. *Perception & Psychophysics*, *59*, 989–1003.

Dixon, W. J., & Mood, A. M. (1948). A method for obtaining and analyzing sensitivity data. *Journal of the American Statistical Association, 43*, 109–126.

García-Pérez, M. A. (1998). Forced-choice staircases with fixed stepsizes: asymptotic and small-sample properties. *Vision Research, 38*, 1861–1881.

Hall, J. L. (1968). Maximum-likelihood sequential procedure for estimation of psychometric functions [abstract]. *Journal of the Acoustical Society of America, 44*, 370.

Hall, J. L. (1981). Hybrid adaptive procedure for estimation of psychometric functions. *The Journal of the Acoustical Society of America, 69*, 1763–1769.

Kaernbach, C. (1991). Simple adaptive testing with the weighted up/down method. *Perception & Psychophysics, 49*, 227–229.

King-Smith, P. E., Grigsby, S. S., Vingrys, A. J., Benes, S. C., & Supowit, A. (1994). Efficient and unbiased modifications of the QUEST threshold method: Theory, simulations, experimental evaluation, and practical implementation. *Vision Research, 34*, 885–912.

King-Smith, P. E., & Rose, D. (1997). Principles of an adaptive method for measuring the slope of the psychometric function. *Vision Research, 37*, 1595–1604.

Kontsevich, L. L., & Tyler, C. W. (1999). Bayesian adaptive estimation of psychometric slope and threshold. *Vision Research, 39*, 2729–2737.

Pelli, D. G. (1987). The ideal psychometric procedure. *Investigative Ophthalmology and Visual Science, 28*(Suppl), 366.

Pentland, A. (1980). Maximum likelihood estimation: the best PEST. *Perception & Psychophysics, 28*, 377–379.

Taylor, M. M., & Creelman, C. D. (1967). PEST: Efficient estimates on probability functions. *The Journal of the Acoustical Society of America, 41*, 782–787.

Watson, A. B., & Pelli, D. G. (1983). QUEST: A Bayesian adaptive psychometric method. *Perception & Psychophysics, 33*, 113–120.

Watt, R. J., & Andrews, D. P. (1981). APE: Adaptive Probit Estimation of Psychometric Functions. *Current Psychological Reviews, 1*, 205–214.

Wetherill, G. B., & Levitt, H. (1965). Sequential estimation of points on a psychometric function. *The British Journal of Mathematical and Statistical Psychology, 18*, 1–10.

CHAPTER

6

Signal Detection Measures

6.1 INTRODUCTION

6.1.1 What is Signal Detection Theory (SDT)?

In performance-based psychophysical tasks there are many situations in which proportion correct (*Pc*) is inappropriate, uninformative, or invalid. In this chapter we will examine some of these situations and describe a popular alternative measure termed d' ("d-prime"). d' is a measure derived from a branch of psychophysics known as signal detection theory (SDT). In Chapter 4 Section 4.3.1.2, SDT was introduced as one of the models of how perceptual decisions were made in a forced-choice task. The SDT model attempted to explain the shape of the psychometric function relating *Pc* to stimulus magnitude. It was argued that the presence of internal noise, or uncertainty, led to stimuli being represented in the brain not by a single point along a sensory continuum, but as a random sample drawn from a distribution with a mean and a variance. SDT is therefore a theory of how observers make perceptual decisions, given that the stimuli are represented stochastically (or probabilistically) inside the brain. The purpose of this chapter is to discuss why d' is a useful measure of performance, to describe the Palamedes routines that convert conventional measures of performance such as *Pc* into d', and to explain the theory behind those conversions.

SDT is a large topic and it is impossible to do it justice in a single chapter of an introductory book on psychophysics. There are a number of excellent books and articles on SDT (see Further Reading at the end of the chapter) that cover a wider range of material and provide a more in-depth treatment than is possible here. In particular, Macmillan and Creelman's (2005) comprehensive treatment of SDT is strongly recommended as an accompaniment to this chapter. Although our chapter is modest by comparison to textbooks specializing in SDT, it nevertheless aims to do more than just scratch the surface of the topic. The Palamedes SDT routines described in Section A of the chapter are generous in terms of the number of psychophysical procedures they cover. Section A is intended to familiarize readers with the basic concepts involved and the practical tools necessary for converting psychophysical measurements into d's, and *vice versa*, without the need to understand the underlying theory. The theory is provided in Section B, and we have attempted to make it as accessible as possible.

6.1.2 A Recap on Some Terminology: N, *m* and M

As elsewhere in this book, we use the term "forced-choice" for any task in which observers are required on each trial to make a forced-choice response, irrespective of the number of stimuli or stimulus alternatives presented on the trial. We denote the number of stimuli presented during a trial as *N*. *m* refers to the number of response choices available to the observer, and as we saw in Chapter 4 this determines the expected chance performance or guessing rate of the task, calculated as $1/m$. Remember that m and N are not always the same. For example, in a yes/no task N is 1 because only one of two stimulus states is presented per trial, target-present or target-absent, but m is 2 because there are two choices of response – "yes" or "no." In a same-different task N can be 2 or 4, depending on whether the Same and Different pairs are presented on separate trials or together in the same trial. When the pairs are presented on separate trials the task for the observer is to respond "same" or "different," whereas when presented together in the same trial the task is to respond "1" or "2," depending on which interval contains the Different (or Same) pair. However, for both varieties of same-different, m is 2.

For the purposes of SDT the third parameter described in Chapter 2, M, is especially important. M is the "number of stimulus alternatives presented per trial." In the $N = 4$ same-different task described above, M is 2, as two stimulus alternatives are presented per trial – the Same pair and the Different pair. Although m is also 2 for this task, m and M are not always the same. For example, with the yes/no task m is 2 but M is 1, since although there are two available response choices, only one stimulus alternative is presented on a trial, either "target present" or "target absent." As we stated in Chapter 2, forced-choice tasks in this book are prefixed with the value of M. Table 6.1 summarizes the relationship between N, m, and M for the main varieties of task discussed in this chapter.

TABLE 6.1 Relationship between M, N, and m for the psychophysical tasks discussed in the chapter

Task	**Acronym prefixed by number of stimulus alternatives per trial M**	**Number of stimuli per trial N**	**Number of response options per trial m**
Yes/no	1AFC	1	2
Symmetric single-alternative	1AFC	1	2
Standard two alternative forced-choice	2AFC	2	2
Single alternative same-different	1AFC	2	2
Two alternative same-different	2AFC	4	2
Two alternative match-to-sample	2AFC	3	2
Three alternative oddity	3AFC	3	3
Standard M-alternative forced-choice	M-AFC	M	M
M-alternative match-to-sample	M-AFC	$M+1$	M
M-alternative oddity	M-AFC	M	M

One could argue that because this chapter deals exclusively with forced-choice tasks, prefixing every task with the acronym AFC is unnecessary because it is redundant, and that instead AFC should be used sparingly as in standard SDT texts (e.g., Macmillan & Creelman, 2005). However, in a book dealing with psychophysical procedures that are not all forced-choice, the acronym helps make explicit those that are.

Recall, also, that for most tasks the stimulus alternatives can be presented in spatial or temporal order, and hence be denoted as AFC or IFC. AFC, however, is the generic acronym, so we have adopted this as the default and used IFC only when specifically referring to tasks in which the stimulus alternatives are presented in temporal order. From the point of view of signal detection theory, however, the two acronyms are interchangeable.

6.1.3 Why Measure d'?

Suppose we wanted to compare the results of two texture-segregation experiments, one that employed a standard 4AFC task and the other a standard 2AFC task. The stimuli employed in texture segregation experiments typically consist of a "target" texture embedded in a "background" texture. The target and background differ in some textural property, such as the average orientation or average size of the texture elements. In the popular 4AFC version, the target is embedded in one of four quadrants of the background texture, and on each trial the observer selects the quadrant containing the target. In the 2AFC version, the embedded target is typically positioned on one or other side of the fixation point. In the 4AFC task proportion correct (*Pc*) would be expected to range from 0.25–1, since the guessing rate $1/m$ is 0.25. For the 2AFC version *Pc* would be expected to range from 0.5–1, since the guessing rate is 0.5. Yet, presumably, the underlying sensory mechanisms involved in segregating the target from the background are the same for both the 2AFC and 4AFC tasks, especially if the target locations are arranged equidistant from fixation in order to stimulate equivalent visual mechanisms. Thus, any differences in performance between the two tasks, which will be most prominent when performance is close-to-chance, are unlikely to be due to differences in the observer's sensitivity to the stimuli, but more probably due to differences in the uncertainty of the target location. Put another way, with four possible target locations the observer is more likely to make a mistake than with two target locations, all else being equal. And the more possible target locations, the more mistakes the observer will likely make. One reason for using d' is that it can remove, or take into account, the effects of target location uncertainty, providing a measure of performance that is procedure-free. In other words, for M-AFC tasks, d' may equate performance across M. We stress "may" because it is ultimately an empirical, not a theoretical, question as to whether d' does equate performance across M, and there are some situations where it has been shown not to do so (e.g., Yeshurun, Carrasco, & Maloney, 2008).

Although the most popular value of M in forced-choice tasks is 2 (either 2AFC or 2IFC), M can be much higher. For example, one of the authors once conducted experiments in which observers were required to choose a column of pixels with the highest average intensity from 256 columns (see Kingdom, Moulden, & Hall (1987) for details). So M for this task was 256!

Another reason for using d' is that in some situations it removes the inherently sigmoidal- or bow-shape of the psychometric function when plotted in terms of Pc. Again, however, it is an empirical question as to whether d' does linearize the psychometric function, and in theory it will only do so if the relationship between sensory magnitude and stimulus magnitude is linear, and the internal noise levels remain constant with stimulus magnitude. However, if these constraints are satisfied Figure 6.1 illustrates how d's equate performance across M and linearize the psychometric function. The figure shows hypothetical Pcs as a function of stimulus level for a standard 2AFC, 4AFC, and 8AFC task, where m is respectively 2, 4, and 8 (see above), under the assumption that observer sensitivity for each stimulus magnitude is the same across tasks. Note that as m increases the functions extend downwards, since chance performance, $1/m$, decreases. When the data are replotted as d's, however, the functions straighten and superimpose.

The linearizing effect of d' is particularly useful when one wants to correlate psychophysical performance with some other dimension. For example, suppose one wanted to study the effects of age on the ability to discriminate the speed of moving vehicles in seniors, using movies of moving vehicles as test stimuli. If insufficient time was available to test each observer with enough stimuli to derive full psychometric functions relating Pc to speed difference, a single Pc at a particular speed

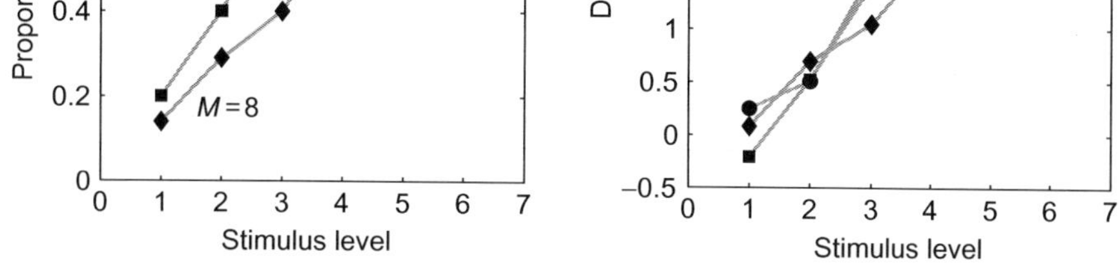

FIGURE 6.1 (a) Hypothetical Pc (proportion correct) data for an $M = 2$, $M = 4$, and $M = 8$ forced-choice task. (b) The same data plotted as d's.

difference might have to suffice. Given the likely range of performance across observers of the same age and across ages, converting the *Pc*s into *d*′s may well linearize the data and render the measured (likely negative) correlation between performance and age more valid. Moreover, correlation is not the only type of statistical manipulation potentially benefitting from converting *Pc*s into *d*′s. Parametric tests such as t-tests and Analysis-of-Variance (ANOVA), which are often employed to make inferences about group differences, are only valid if the data within each group are normally distributed. Converting *Pc*s to *d*′s may turn out to satisfy this requirement.

A third, and for many investigators the most important, reason for using *d*′s is that certain types of psychophysical task are prone to the effects of observer bias. In Chapters 2 and 3 we noted for example, that in the yes/no task observers tend to be biased towards responding "yes" or "no," irrespective of their underlying sensitivity to the target. As we shall see, the greater the bias, the smaller the expected *Pc*, making *Pc* an invalid measure of sensitivity. *d*′ takes into account the effects of bias, thus providing a relatively "bias-free" measure of performance.

6.2 SECTION A: PRACTICE

6.2.1 Signal Detection Theory with Palamedes

Palamedes contains a large number of routines for performing signal detection computations. To understand the convention for the names of the routines, consider the following example: **`PAL_SDT_2AFCmatchSample_DiffMod_PCtoDP`**. The routine is identifiable as part of the SDT package by the prefix **`PAL_SDT`**. The term **`2AFCmatchSample`** identifies the task, i.e., match-to-sample, and the value of *M*, i.e., 2. The generic acronym AFC is used in all routines. The term **`DiffMod`** specifies the particular model for the computation; in this case it means a differencing model. By "model" we mean the particular strategy that the observer is assumed to adopt when performing the task. This is only relevant to those tasks for which there is more than one possible strategy. Finally **`PCtoDP`** specifies the actual computation, in this case the conversion of *Pc* to *d*′. This last term is invariably of the form XtoY, where X is the principle input argument(s) and Y the principle output argument(s). However, the routine may require additional input arguments and may output additional arguments. The abbreviations used for the XtoY term are **`PC`** for *Pc*, **`DP`** for *d*′, **`PHF`** for proportion hits and proportion false alarms, and **`PH`** for proportion hits. If the required input to a routine is either **`PC`** or **`DP`**, the data can be entered as a scalar, vector or matrix. If the input is **`PHF`** the data must be an $m \times 2$ (rows × columns) matrix with a minimum of one row; one column is for the proportion of hits and the other column for the corresponding proportion of false alarms.

There are two important assumptions underlying all the routines described in this chapter. The first concerns the stimuli. We assume that all the stimulus alternatives are presented the same number of times in each experiment. Thus, in a yes/no task we assume that there are as many target-present as target-absent trials, and in a same-different task as many Same pairs as Different pairs. The second assumption is that the observer's internal noise variance is the same across stimulus alternatives. This is the "default" assumption for most SDT tasks. There are procedures for determining from data whether this assumption is violated, but they will not be dealt with here and the interested reader is referred to Macmillan and Creelman (2005) for the necessary details.

6.2.2 Converting *Pc* to *d'* for Unbiased M-AFC Tasks

We begin with the routine that deals with the standard *M*-AFC task, where *M* is any value greater than 1. For this class of task the three variables, *N*, *M*, and *m*, are equal. Although tables exist for converting *Pc* to *d'* for a range of *M* (Elliot, 1964; Macmillan & Creelman, 2005) the Palamedes routines work for any value of *M* and are simple to implement.

In an *M*-AFC task, one of the alternatives on each trial contains the target, while the remaining *M*-1 alternatives contain no target. If the observer selects the alternative containing the target their response is scored "correct," otherwise it is "incorrect," and *Pc* is calculated as the proportion of trials in which the observer is scored correct.

The two routines for standard *M*-AFC are **PAL_SDT_MAFC_DPtoPC**, which converts *d'* to *Pc*, and **PAL_SDT_MAFC_PCtoDP**, which converts *Pc* to *d'*. The routines make an important assumption in addition to those described in the previous section. The assumption is that the observer is not biased to respond to any one alternative/interval more than any other. If the assumption is not true and the observer is biased, then the estimates of *d'* will not be close to the "true" values.

Each routine takes two arguments. The first is a scalar, vector or matrix of the measure to be converted (*d'* or *Pc*), and the second is the value of *M* where $M > 1$. Typically one wants to convert *Pc*s into *d'*s, so try filling a vector named **PropCorr** with an array of *Pc*s as follows:

```
>> PropCorr = [.3:.1:.9];
```

To convert the array to *d'*s for, say, a 3AFC task, type and execute:

```
>> DP = PAL_SDT_MAFC_PCtoDP(PropCorr,3)
```

The array returned is:

```
DP =
-0.1207 0.2288 0.8852 1.6524 2.2302
```

Note that the first value in the array is negative. *d'* is zero when performance is at chance, which for the standard 3AFC task is 0.33, so any *Pc* below 0.33 will produce a negative d'. Try repeating the above with $M = 4$. Note that the first value is now positive, since chance level for a 4AFC task is 0.25. If one sets M to 2 (chance = 0.5) the first *two* d's are negative. One can see from these examples that increasing M for a given Pc increases d'. This is because as M increases so too does the chance that one of the non-target intervals/locations will contain a signal that is by chance greater in magnitude than the interval/location containing the target. In other words, for a given Pc, observer sensitivity is computed to be higher if the task has a large compared to a small M.

Try also converting an array of d's to Pcs using **PAL_SDT_MAFC_DPtoPC**. Note that increasing M for a given d' this time decreases Pc, because the more possible target intervals/ locations, the more likely one of them will, by chance, contain a signal greater than the interval/location containing the target.

6.2.3 Measuring d′ for 1AFC Tasks

6.2.3.1 *d′ from pH and pF*

As discussed in Chapters 2 and 3, 1AFC tasks, especially those that are not symmetric such as the yes/no task, are particularly prone to bias. Remember that with the yes/no task, the observer is required to indicate on each trial whether the target stimulus is present or absent. If the observer adopts a loose criterion, this will result in a bias towards responding "yes," whereas adopting a strict criterion will result in a bias towards responding "no." Both types of bias may occur irrespective of how sensitive the observer is to the stimulus. For this reason, signal detection theory approaches the computation of d' for 1AFC tasks differently from tasks that are assumed to be bias-free. Rather than use Pc, the responses from a 1AFC task are divided into two groups: the target-*present* trials in which the observer responds "yes" (i.e., correctly) and the target-*absent* trials in which the subject responds "yes" (i.e., incorrectly). The former responses are commonly termed "hits," the latter "false alarms." The proportion of target-present trials that are hits is given here by pH, and the proportion of target-absent trials that are false alarms, pF. Note that the overall Pc is given by $[pH + (1 - pF)]/2$, since $1 - pF$ gives the proportion of target-absent trials in which the observer responds "no," i.e., also correctly. The two measures pH and pF can be used not only to calculate d', but also to calculate the bias towards responding "yes" or "no." The calculations are explained in Section B.

The Palamedes routine that converts pH and pF to d', as well as to two measures of bias, is **PAL_SDT_1AFC_PHFtoDP**. The input argument can either be a pre-named $m \times 2$ matrix of pH and pF values, or the raw values themselves. There are four output arguments: d'; two measures of bias termed C and $\ln\beta$; and overall Pc.

Suppose we want to input just a single pair of raw pH and pF values. Type and execute the following, and remember to place the square brackets around the two values so that they are entered as a matrix:

```
>> [dp C lnB Pc] = PAL_SDT_1AFC_PHFtoDP([0.6 0.1])
```

The output should be:

```
dp =
1.5349
C =
0.5141
lnB =
0.7891
Pc =
0.7500
```

The criterion C can range from negative to positive, with negative values indicating a bias towards "yes" and positive values a bias towards "no." The criterion measure lnβ shows the same pattern as C. The positive values of $C = 0.51$ and ln$\beta = 0.78$ in the above example are indicative of a relatively strict criterion, that is a bias towards responding "no."

To explore the relationship between d', bias, pH, and pF, one can also use the reverse routine **PAL_SDT_1AFC_DPtoPHF**. For example, create a vector named **dprime** filled with a ×5 array of 2s, and a vector named **criterion** with values −1, −0.5, 0, 0.5, and 1. Then type and execute:

```
>> pHF = PAL_SDT_1AFC_DPtoPHF(dprime,criterion)
```

The output should be:

```
pHF =
0.9772 0.5000
0.9332 0.3085
0.8413 0.1587
0.6915 0.0668
0.5000 0.0228
```

The first column gives pH, the second pF. Note that as C increases (loose to strict criterion) both the number of hits *and* the number of false alarms decreases. Figure 6.2a shows the relationship between pH and pF as a function of C for three values of d'. As one travels along each of the curves from left to right, C is decreasing, resulting in an increase in both pH and pF. The relationship between pH and pF is known as a receiver operating characteristic, or ROC. The ROC in Figure 6.2a is hypothetical, but ROCs can be generated from experiments in which the responses are not binary options such as "yes" or "no," but ratings, for example 1 to 5, as to how confident one is that the target is present. Currently Palamedes does not provide routines for analyzing rating-scale data, but the method, along with the value of ROCs for testing the equal-variance assumption mentioned above, is described in Macmillan and Creelman (2005).

Figure 6.2.b shows why Pc is not a good measure of performance when there is bias. Assuming that the "true" observer's sensitivity is given by d', one can see that

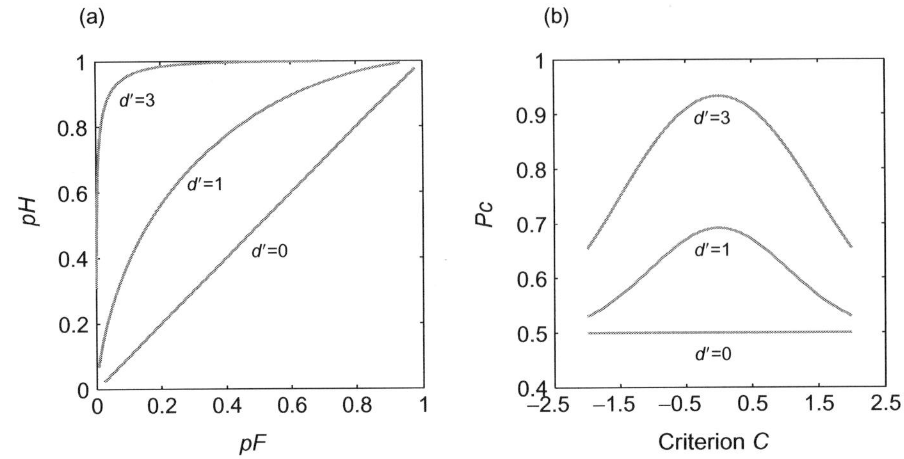

FIGURE 6.2 (a) Hypothetical receiver operating characteristics, or ROCs, for three d's. Note that pH (proportion of hits) is plotted against pF (proportion of false alarms). As the criterion C decreases, one moves along each curve from left to right. (b) the effect of changing C on the overall Pc (proportion correct) for the same d's as in (a). Note that all three curves peak when $C = 0$.

Pc varies considerably with criterion C. Only if there is no bias ($C = 0$) is Pc a valid measure of performance. A zero-bias assumption may sometimes be reasonable with "symmetric" 1AFC tasks, such as the 1AFC orientation discrimination experiment discussed in Chapter 3. However, some researchers argue that even with symmetric 1AFC tasks the data should be analyzed under the presumption that bias might occur. If it turns out there is no bias then nothing is lost, but if bias is found to occur it is taken into account – a win-win situation.

How, then, do we convert the responses from the orientation discrimination experiment into pH and pF? The answer is to classify the responses in a way analogous to that of the yes/no experiment. For the orientation discrimination experiment this means classifying a "left-oblique" response as a "hit" when the stimulus is left-oblique, and as a "false alarm" when the stimulus is right-oblique. pH is then the proportion of "left-oblique" responses for the left-oblique stimuli, and pF the proportion of "left-oblique" responses for the right-oblique stimuli. Note that pH and pF defined in this way are sufficient to describe all the responses in the experiment i.e., including the "right-oblique" responses. The proportion of times the observer responds "right-oblique" is $1 - pH$ for the left-oblique stimulus trials and $1 - pF$ for the right-oblique stimulus trials. Note also that as with yes/no, overall Pc is given by $[pH + (1 - pF)]/2$. Thus, if the observer in the orientation discrimination experiment is biased towards responding "left-oblique," both pH and pF, as defined above, will tend to be relatively high, and by comparing the two in the same way as with the yes/no task the bias can be taken into account and a valid measure of sensitivity calculated.

6.2.3.2 *1AFC Demonstration Programs*

The program **PAL_SDT_1AFC_PHFtoDP_Demo** illustrates how the routines for 1AFC tasks can be incorporated into a program that generates a more user-friendly output of *d′* and criterion measures. When executed, the program prompts you as follows:

```
Enter a matrix of proportion Hits and False Alarms
```

You must enter arrays of raw values. An example input matrix of *pH* and *pF* values would be:

```
[0.6 0.2; 0.7 0.2; 0.8 0.2]
```

The output should be:

pH	pF	Dprime	propCorr	Crit C	lnBeta
0.6000	0.2000	1.0950	0.7000	0.2941	0.3221
0.7000	0.2000	1.3660	0.7500	0.1586	0.2167
0.8000	0.2000	1.6832	0.8000	-0.0000	-0.0000

The inverse routine **PAL_SDT_1AFC_DPtoPHF_Demo** operates similarly. You are prompted for two vectors of numbers. Try entering the following values, then execute:

```
Enter a vector of Dprime values [1 2 3]
Enter a vector of Criterion C values [0.5 0.5 0.5]
```

The output should be:

dprime	critC	pH	pF	pCorr
1.0000	0.5000	0.5000	0.1587	0.6707
2.0000	0.5000	0.6915	0.0668	0.8123
3.0000	0.5000	0.8413	0.0228	0.9093

6.2.4 Measuring *d′* for 2AFC Tasks with Observer Bias

Although the inherent symmetry of 2AFC tasks makes them less susceptible to bias than the yes/no task, a bias towards responding to one alternative/interval more than the other may still occur, and if it does occur *Pc* becomes an invalid measure of sensitivity. As with symmetric 1AFC tasks, some researchers prefer not to hedge their bets with 2AFC and analyze the data on the presumption that bias might have occurred.

To take into account bias in 2AFC tasks, the observer's responses need to be classified as hits and false alarms, as with the symmetric 1AFC task. Let the response be "1" or "2" depending on the alternative perceived to contain the target. A "1" response is designated as a "hit" when the target is present in the first alternative/interval, and as a "false alarm" when the target is present in the second alternative/interval. Thus *pH* is the proportion of "1" responses for targets presented in the first alternative/interval and *pF* the proportion of "1" responses for targets presented in the second alternative/interval. Note that, as with 1AFC tasks, *pH* and *pF* defined in this way are sufficient to describe the full pattern of responses. Thus 1-*pH* is the proportion of "2" responses for targets presented in the first alternative/interval and $1 - pF$ the proportion of "2" responses for targets presented in the second alternative/interval. Note also that, as with 1AFC tasks, overall *Pc* is given by $[pH + (1 - pF)]/2$.

Palamedes has two routines for the standard 2AFC task when the input arguments are *pH* and *pF*: **`PAL_SDT_2AFC_DPtoPHF`** and **`PAL_SDT_2AFC_PHFtoDP`**. The input and output arguments correspond to those for the 1AFC routines. Remember that one can also use **`PAL_SDT_MAFC_PCtoDP`** and **`PAL_SDT_MAFC_DPtoPC`** for 2AFC tasks (by inputting data in the form of *Pc* and setting the argument *M* to 2), but only if one is happy to assume that the observer is unbiased.

What is the expected relationship between performance in a 1AFC and 2AFC task? One can use **`PAL_SDT_1AFC_PHFtoDP`** and **`PAL_SDT_2AFC_PHFtoDP`** to find out. Try the following. Input the same pair of *pH* and *pF* values and the same value of the criterion to both routines. Take the ratio of the resulting 1AFC to 2AFC *d*'s. The result should be $\sqrt{2}$. The $\sqrt{2}$ relationship between *d*'s for 1AFC and 2AFC is often emphasized in expositions of SDT, but one must be careful with its interpretation. It is tempting to suppose that if one performed a 1AFC task and a 2AFC task using the same stimulus magnitudes, the computed *d*'s would likely come out in a ratio of $\sqrt{2}$. In fact, the *d*'s would likely be very similar. Remember that d' is a measure of sensitivity that is ostensibly independent of the method used to obtain it (although be reminded of the cautionary note from Yeshurun et al., 2008). The likely difference between the two tasks will be in *Pc*, not d'. As Figure 6.3 demonstrates, the same d' predicts different *Pc*s for 1AFC and 2AFC. Put another way, observers will typically find a 1AFC task more difficult than a 2AFC task for the same stimulus magnitudes. This is because there is more information in a 2AFC compared to 1AFC trial.

6.2.5 Measuring d' for Same-Different Tasks

In Chapters 2 and 3 we described the class of psychophysical task termed "same-different." One reason for using same-different tasks is that the observer is not required to know the basis on which the discriminands differ. There are two main varieties of same-different task. In the 1AFC version only one pair, Same *or* Different, is presented on a trial, and the observer has to decide "same" or "different." The pair can be presented either together on the display or in temporal order.

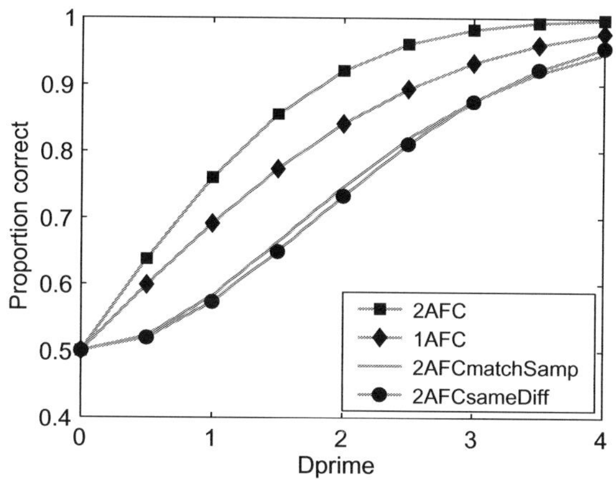

FIGURE 6.3 Example output from **PAL_SDT_DPtoPCcomparison_Demo**. When using **PAL_SDT_PCtoDPcomparison_Demo** do not enter any 1s as this will result in a d' of infinity which cannot be plotted. Instead an example vector input could be **[.5:.025:0.98]**.

In the 2AFC version the Same *and* Different pairs are both presented on a trial (either together on the display or in temporal order), and the observer chooses the alternative/interval containing the Different (or the Same) pair. The 2AFC same-different task is probably the more popular of the two versions in vision experiments, because it is less prone to bias.

6.2.5.1 *d′ for 2AFC Same-Different*

The Palamedes routines for the 2AFC same-different task are **PAL_SDT_2AFCsameDiff_DPtoPC** and **PAL_SDT_2AFCsameDiff_PCtoDP**. Both routines assume an unbiased observer that adopts the strategy of selecting the pair with the greater (or smaller) absolute perceived difference. The routines implement the equations in Macmillan, Kaplan, and Creelman (1977) for a "4IAX" same-different task, where 4IAX denotes that the four stimuli are presented in temporal order, the more typical scenario in an auditory experiment. Both of the Palamedes routines take a single argument (d' or Pc) and output a single argument (Pc or d'). The input arguments may be scalars, vectors or matrices.

6.2.5.2 *d′ for 1AFC Same-Different*

For same-different tasks where only one pair, Same or Different, is presented in a trial, Macmillan and Creelman (2005) argue that observers typically adopt one of two strategies: the "independent observation" or "differencing" strategy (note that Macmillan and Creelman occasionally refer to the 1AFC same-different task as 2IAX or AX. The first acronym denotes that the two stimuli are presented in different temporal intervals).

Suppose that during a session there are only two stimuli: S_1 and S_2. On each trial the observer is presented with one of four possible combinations: $<S_1S_1>$, $<S_2S_2>$, $<S_1S_2>$, or $<S_2S_1>$. Macmillan and Creelman argue that the most likely strategy in this scenario is that the observer independently assesses the likelihood that each stimulus in a pair is either S_1 or S_2. The decision "different" is made when the joint likelihood of the pair being S_1 and S_2 exceeds the observer's criterion. This is the independent observation strategy.

The differencing strategy is less optimal, but under some circumstances the more likely to be adopted. As with the strategy assumed for the 2AFC same-different task described above, the decision rule is based on the perceived *difference* between the two stimuli in each pair. The observer responds "different" when the absolute perceived difference between the two stimuli exceeds the criterion. According to Macmillan and Creelman, the differencing strategy is more likely to be adopted when many different stimuli are presented during a sesssion, termed a "roving" experiment. For example, suppose that one wished to compare the detectability of four types of color manipulation applied to images of natural scenes. Let the four manipulations be shifts in average color towards either red, green, blue or yellow. On each trial observers are presented either with two identical natural-scene images (the Same pair) or two images in which the average color of one of the pair was shifted towards one of the four (randomly selected) colors (the Different pair). It would be difficult for observers to independently assess the likelihood that each member of a pair had been subject to a particular color shift, because there are four possible types of color shift. The more likely strategy in this situation would be that observers assess the difference in color between the images in each pair and base their decision accordingly.

6.2.5.2.1 *d′* for 1AFC Same-Different: Independent Observation Model

The Palamedes routines for the 1AFC same-different task that assumes an independent-observation model are **PAL_SDT_1AFCsameDiff_IndMod_PHFtoDP** and **PAL_SDT_1AFCsameDiff_IndMod_DPtoPHF**. The first routine takes two arguments, a m $\times$ 2 matrix of *pH*s and *pF*s, and outputs two arguments: *d′* and criterion *C*. The second routine performs the reverse operation. For example, try inputting the same matrix of *pH*s and *pF*s as for the basic 1AFC task described earlier, i.e.:

```
PHF = [0.6 0.2; 0.7 0.2; 0.8 0.2]
```

then type and execute:

```
>>[dp C] = PAL_SDT_1AFCsameDiff_IndMod_PHFtoDP(PHF)
```

The output should be three *d′* and three criterion *C* values. Compare these with those obtained using **PAL_SDT_1AFC_PHFtoDP**. Try also computing *d′*s for a bias-free version of the same-different task by setting *pF* equal to $1 - pH$. You will see that the resulting *d′*s under the independent observation model are the same as those for the 2AFC same-different task ($Pc = pH$), which assumes a differencing strategy.

6.2.5.2.2 d' for 1AFC Same-Different Tasks: Differencing Model

The Palamedes routines for the 1AFC same-different task assuming the differencing model are **PAL_SDT_1AFCsameDiff_DiffMod_PHFtoDP** and **PAL_SDT_1AFCsameDiff_DiffMod_DPtoPHF**. They are implemented in the same way as the routines for the independent observer model. However, they return a different measure of bias termed k (Macmillan & Creelman, 2005). Unlike C, k is not zero when the observer is unbiased.

Consider the following. For a given pH and pF, would you expect d' to be larger or smaller for the differencing compared to the independent observer model? Try various pH and pF combinations to test your predictions.

6.2.6 Measuring d' for Match-to-Sample Tasks

In a match-to-sample task, the observer is presented with a "Sample" stimulus followed by two or more "Match" stimuli, one of which is the same as the Sample – the one the observer must choose. Match-to-sample procedures are particularly popular in animal research, research into children's perception, and studies of cognitive vision (see Chapter 3). As with the same-different task, one advantage of match-to-sample over standard M-AFC is that the observer need not know the basis on which the discriminands differ. The minimum number of stimuli per repeat trial in a match-to-sample task is three (one Sample; two Match), and this is undoubtedly the most popular design. With two Match stimuli the task is 2AFC according to our naming system. Macmillan and Creelman (2005) refer to the task as ABX.

6.2.6.1 d' for 2AFC Match-to-Sample

Macmillan and Creelman argue that for the ABX task, observers may adopt either independent observation or differencing strategies, the latter more likely in "roving" experiments where a number of different stimulus pairs are presented during a session. The independent observation strategy is analogous to that for the same-different task. When adopting the differencing strategy the observer selects the Match that is perceived to be *least* different from the Sample. Palamedes provides eight routines for the 2AFC match-to-sample task:

PAL_SDT_2AFCmatchSample_DiffMod_PCtoDP
PAL_SDT_2AFCmatchSample_DiffMod_DPtoPC
PAL_SDT_2AFCmatchSample_DiffMod_PHFtoDP
PAL_SDT_2AFCmatchSample_DiffMod_DPtoPHF
PAL_SDT_2AFCmatchSample_IndMod_PCtoDP
PAL_SDT_2AFCmatchSample_IndMod_DPtoPC

PAL_SDT_2AFCmatchSample_IndMod_PHFtoDP
PAL_SDT_2AFCmatchSample_indMod_DPtoPHF

The routines use the same input and output arguments as the same-different routines. Given that observers might be biased towards choosing one Match alternative over the other, it is recommended to use the routines that take *pH* and *pF* rather than *Pc* as arguments, unless there is good reason to assume the observer is unbiased.

6.2.6.2 *d′ for M-AFC Match-to-Sample*

For $M > 2$ match-to-sample tasks, Palamedes has two routines:

PAL_SDT_MAFCmatchSample_DPtoPC and
PAL_SDT_MAFCmatchSample_PCtoDP.

Both routines assume that the observer is unbiased and adopts a differencing strategy. In keeping with other Palamedes SDT routines, each routine takes two input arguments: a scalar, vector or matrix of *Pc*s or *d′*s, and a value of *M*. The output arguments are *d′*s or *Pc*s. The reader will find that the **_DPtoPC** routine is slower to execute than other SDT routines. This is because the calculations are implemented by Monte Carlo simulation using a very large number of trials. The reverse routine, **_PctoDP**, is even slower as it performs an iterative search based on the forward routine. As a result, the routine may take minutes to execute depending on the speed of the computer and the number of input *Pc* values.

6.2.7 Measuring *d′* for M-AFC Oddity Tasks

In an oddity task, often termed an "odd-man-out" task, the observer is presented with an array of stimuli, all but one of which are the same, and chooses the stimulus that is different, in other words the "oddity." As with the same-different and match-to-sample tasks, the observer in an oddity task does not need to know the basis on which the stimuli differ. Probably the most popular form of oddity task is the one using the minimum number of alternatives per trial, which is three, and for this reason sometimes termed the "triangular" method. However, the principle extends to any *M*. One likely strategy in an oddity task is that observers select the alternative that is most different from the mean of all the alternatives, another instance of a differencing strategy. Craven (1992) has provided a table for converting *Pc* into *d′* for oddity tasks with various *M*, assuming this strategy and an unbiased observer. Palamedes provides two routines that perform the same computations as those described in Craven (1992), but for any *M*: **PAL_SDT_MAFCoddity_DPtoPC** and **PAL_SDT_MAFCoddity_PCtoDP**. As elsewhere, each routine takes two arguments: a scalar, vector or matrix of *Pc*s or *d′*s, and the value of *M*. The output arguments are *d′*s or *Pc*s. As with the *M*-AFC match-to-sample routines described in the previous section, the *M*-AFC oddity routines are slow to execute as they also employ Monte Carlo simulation.

6.2.8 Estimating Pc_{max} with Observer Bias

As we have argued above, *Pc* is not a valid measure of performance for any of the procedures described if there is a significant amount of observer bias. However, it is possible to obtain an estimate of the *Pc* that would be expected if the observer were not biased. This is termed Pc_{max} (or Pc_{unb}), because *Pc* reaches a theoretical maximum when there is no bias (e.g., see Figure 6.2b for the 1AFC task). One can think of Pc_{max} as an unbiased estimate of *Pc*. Estimating Pc_{max} is straightforward with Palamedes, provided one has available a measure of the criterion that is zero when the observer is unbiased, as with the routines that compute the criterion measure *C*. To obtain Pc_{max} one inputs *pH* and *pF* into the relevant routine (i.e., one ending in **_PHFtoDP**) to obtain d' and a measure of *C*, and then use the reverse routine (the same routine ending in **_DPtoPHF**) to convert back to *pH* and *pF*, using as the input argument a zero value for *C*. Pc_{max} is then equal to the output *pH*.

Take the following example. Suppose you want to estimate Pc_{max} for a 2AFC match-to-sample task assuming a differencing strategy. Let $pH = 0.8$ and $pF = 0.6$. One can glean from these values that the observer is biased towards the alternative for which a correct response is classified as a "hit," since the number of false alarms exceeds $1 - 0.8$ i.e. 0.2. Recall also that *Pc* is given by $[pH + (1 - pF)]/2$, which for this example is 0.6. If the values of *pH* and *pF* are input to **PAL_SDT_2AFCmatchSample_DiffMod_PHFtoDP**, the routine returns a d' of 1.2137 and a criterion *C* of -0.5475. If one now inputs the same d' to **PAL_SDT_2AFCmatchSample_DiffMod_PHFtoDP**, but with *C* set to zero, the outputs are $pH = 0.6157$ and $pF = 0.3843$. Thus, Pc_{max} is 0.6157. Pc_{max} is only slightly higher than the actual *Pc* because the bias in the example is not particularly strong.

6.2.9 Comparing d's and *Pc*s across Different Tasks

Two scripts are provided by Palamedes that demonstrate the differences between the computed d's and *Pc*s for a variety of tasks: **PAL_SDT_DPtoPCcomparison_Demo** and **PAL_SDT_PCtoDPcomparison_Demo**. The tasks compared are 1AFC, standard 2AFC, 2AFC same-different, and 2AFC match-to-sample. The standard 2AFC, same-different, and match-to-sample tasks assume a differencing strategy, and all tasks assume an unbiased observer. Therefore, for the 1AFC tasks, criterion *C* is set to zero to produce an optimal *Pc*. The scripts prompt you either for d's or *Pc*s. Try the first program:

```
>>PAL_SDT_DPtoPCcomparison_Demo
```

Enter a vector of **Dprime** values and enter:

```
[0:.5:4]
```

The output should look like this:

```
------------Proportion correct-----------

dprime    1AFC     2AFC     2AFCsameDiff    2AFCmatchSamp
0         0.5000   0.5000   0.5000          0.5000
0.5000    0.5987   0.6382   0.5195          0.5223
1.0000    0.6915   0.7602   0.5733          0.5825
1.5000    0.7734   0.8556   0.6495          0.6635
2.0000    0.8413   0.9214   0.7330          0.7468
2.5000    0.8944   0.9615   0.8110          0.8196
3.0000    0.9332   0.9831   0.8753          0.8765
3.5000    0.9599   0.9933   0.9231          0.9178
4.0000    0.9772   0.9977   0.9555          0.9467
```

and a graph will be plotted as in Figure 6.3 shown above.

6.3 SECTION B: THEORY

6.3.1 Relationship Between Z-scores and Probabilities

To understand the theory behind calculations of d' it is necessary to begin with some basics. An important relationship that underpins much of SDT is that between z-values and probabilities. Figure 6.4 shows a "standardized" normal probability distribution. This is a normal distribution in which the abscissa is given in units of standard deviation, or z units. The ordinate in the graph is termed "probability density" and denoted by ϕ. Probability density values are not actual probabilities of z-values, but their relative likelihoods, specifically derivatives or "rates of change" of probabilities. Thus, in order to convert z units, or rather intervals between z units, into probabilities, one has to integrate the values under the curve between z-values. If one integrates the curve between $-\infty$ and some value of z, the result is a value from a distribution termed the cumulative normal. Because the total area under the standard normal distribution is by definition unity, the cumulative normal distribution ranges from 0–1. The cumulative normal gives the probability that a random variable from a standardized normal distribution is less than or equal to z.

The equation for the standardized normal distribution is:

$$\phi(z) = \frac{1}{\sqrt{2\pi}} \exp\left(\frac{-z^2}{2}\right) \tag{6.1}$$

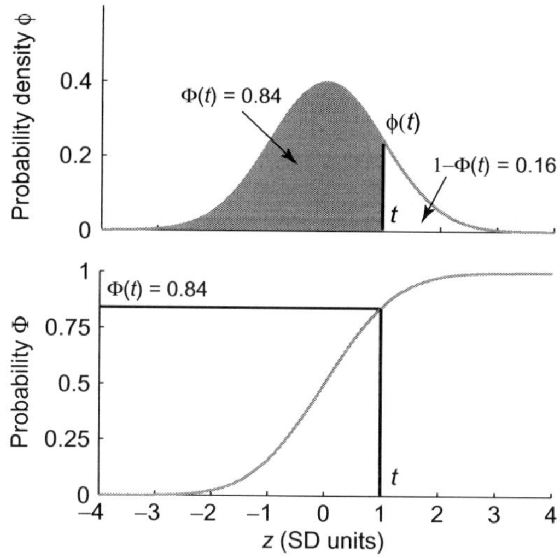

FIGURE 6.4 Relationship between z, probability density ϕ, and cumulative probability Φ. Top: standardized normal distribution. Bottom: the integral of the standardized normal distribution, or cumulative probability distribution. The value of z at point t is 1 in both graphs. In the top graph the height of the distribution at t is denoted by $\phi(t)$ and the area under the curve to the left of t (shown in gray) which has a value of 0.84 denoted as $\Phi(t)$. In the bottom graph $\Phi(t)$ is now a point on the ordinate. In the top graph the white area to the right of t, defined as $1 - \Phi(t)$ has a value 0.16.

and for the cumulative normal:

$$\Phi(z) = 0.5 + 0.5erf(z/\sqrt{2}) \tag{6.2}$$

where *erf* stands for the "error function", which performs the integration. The two values of 0.5 in the equation convert the range of the function to 0–1. The inverse of the cumulative normal, which converts Φ to a z-value is:

$$z(\Phi) = \sqrt{2}erfinv(2\Phi - 1) \tag{6.3}$$

Palamedes contains two routines, **PAL_ZtoP** and **PAL_PtoZ**, which implement, respectively, Equations 6.2 and 6.3. Given that z-values are symmetric around zero, we can state two simple relationships:

$$1 - \Phi(z) = \Phi(-z) \tag{6.4}$$

and

$$-z(\Phi) = z(1 - \Phi) \tag{6.5}$$

You can verify these relationships and the values illustrated in the figure using **PAL_ZtoP** and **PAL_PtoZ**.

6.3.2 Calculation of d' for M-AFC

We begin by describing the theory behind the computation of d' for a standard M-AFC task, where M can be any value greater than 1, and where $M = N = m$. We remind the reader that the calculations described in this section are based on two assumptions, the first that the observer is unbiased and the second that the internal responses to all the stimulus alternatives are normally distributed and of equal variance. Although readers will mainly use the routine **PAL_SDT_MAFC_PCtoDP**, which converts Pcs to d's, it is best to begin with the theory behind its inverse: **PAL_SDT_MAFC_PCtoDP**.

Figure 6.5 shows two standardized normal distributions. One represents the distribution of sensory magnitudes or internal responses to a "blank" interval or location, i.e., one without a target and denoted by "noise alone" or N. The other represents the distribution of sensory magnitudes to the interval/location containing the target, typically denoted in the SDT literature as "signal-plus-noise" or $S + N$. Note, however, that N versus $S + N$ is not the only scenario for which the present analysis is applicable. The principle also extends to the situation in which one interval/location contains stimulus S_1 while the remaining intervals/locations contain stimulus S_2.

Representing the sensory magnitudes of N and $S + N$ as probability distributions means that on any trial the actual sensory magnitudes will be random samples from those distributions. The relative probabilities of particular samples are given by the heights of the distributions at the sample points.

The aim of the observer in the standard forced-choice task is to identify on each trial the alternative containing the target. Let us assume that the observer adopts what is intuitively the optimum strategy: select the alternative with the biggest signal. Try to imagine a strategy that would result in better performance. There isn't one. The rule employed by the observer for selecting the target is usually termed the "decision rule." The question then becomes: how well will the observer do, as measured by Pc,

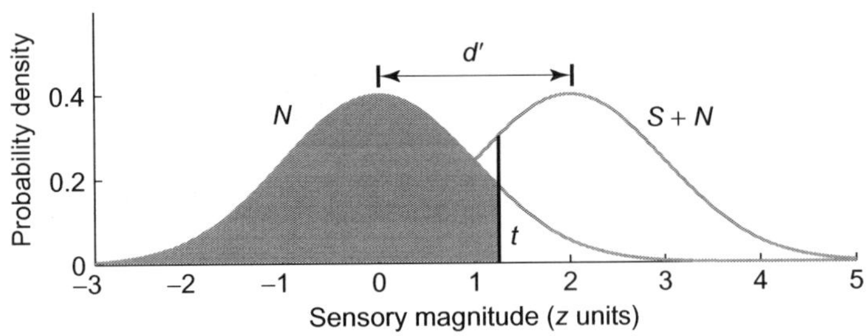

FIGURE 6.5 Calculation of Pc from d'. N = noise; $S + N$ = signal-plus-noise; t is a random variable. See text for details.

when adopting this decision rule? If we make the two assumptions stated above, then the computation of d' turns out to be reasonably straightforward.

One can glean from Figure 6.5 that when there is little overlap between the N and $S + N$ distributions the observer will perform better than when there is a lot of overlap. The reason for this is that, as the N and $S + N$ distributions draw closer together, there is an increasing likelihood that a sample drawn randomly from the N distribution will be greater in magnitude than a sample drawn randomly from the $S + N$ distribution. Each time this happens the observer will make a mistake if adopting the "select the biggest signal" decision rule. If there were no overlap at all between the two distributions, the observer would never make an incorrect decision using this rule, and if the distributions perfectly overlapped the observer would perform at chance. Thus, the degree of overlap between the two distributions is the critical determinant of performance. And because the overlap is governed by two factors, first the separation of the two distributions and second their spread, or σ, one can see that the measure d', which is the separation between the distributions expressed in units of σ, captures the discriminability of N and $S + N$. But how do we calculate the expected Pc, given d' and M?

Suppose that on a given trial the target stimulus has a sensory magnitude given by t in the figure. Remember that t is a random sample, meaning that t will vary between trials, and that the relative probability of a given t is given by the height of the distribution at t. The probability that t will be greater than a random sample from just *one* noise (N) location is given by the gray area to the left of t under the noise distribution. This is simply $\Phi(t)$, since we have (arbitrarily) centered the noise distribution at zero. However, we do not just wish to know the probability that t will be greater than a random sample from just one noise location, but from $M - 1$ noise locations. In other words, we want to know the probability that t will be greater than a random sample from noise location 1 *and* noise location 2 *and* noise location 3 *and* 4 and so on, up to $M - 1$. The "and" term here implies a joint probability, and if we assume that the samples from the different noise locations are independent, this is obtained simply by multiplying the individual probabilities. Since we are muliplying the same thing over again we simply raise the probability to the power of $M - 1$, and hence obtain $\Phi(t)^{M-1}$. However, this still only gives us the probability that *one specific* random sample from the signal distribution, t, will be greater than all random samples from all $M - 1$ noise locations. To obtain the probability that *a random sample t* will be greater than random samples from $M - 1$ noise locations, which gives us our Pc, we need to integrate the above result across all possible values of t. We do this by multiplying $\Phi(t)^{M-1}$ by the height, or relative likelihood of t, which is given by $\phi(t - d')$ (the S distribution is offset from zero by d'), and integrating over all possible values of t. Hence we have:

$$Pc = \int_{-\infty}^{\infty} \phi(t - d') \cdot \Phi(t)^{M-1} dt \tag{6.6}$$

(Green & Swets, 1974; Wickens, 2002). The function **PAL_SDT_MAFC_DPtoPC** implements this equation using the numerical integration function **quadgk** in MATLAB®.

How do we convert a *Pc* into a d' for an $M -$ AFC task, which is our primary aim? Equation 6.6 is not easily invertible, so **PAL_SDT_MAFC_PCtoDP** performs an iterative search using the **fminsearch** function in MATLAB to find that value of d' which, when converted to *Pc* (using **PAL_SDT_MAFC_DPtoPC**), gives the input *Pc*.

6.3.3 Calculation of d′ and Measures of Bias for 1AFC Tasks

6.3.3.1 *Calculation of d′ for 1AFC*

Let us consider the 1AFC task known as yes/no, a task that is particularly prone to bias. Adopting the same scheme for representing the distributions of sensory magnitudes as in the previous section for the standard *M*-AFC task, the situation is illustrated in Figure 6.6. This time, the *N* and $S + N$ distributions are shown separately as the stimuli they represent are presented on separate trials. The gray areas to the right of the vertical criterion line represent sensory magnitudes that the observer deems large enough to warrant a "yes" response. Sensory magnitudes to the left of this line produce a "no" response. The gray area to the right of the criterion in the lower $S + N$ distribution gives the proportion of target-present

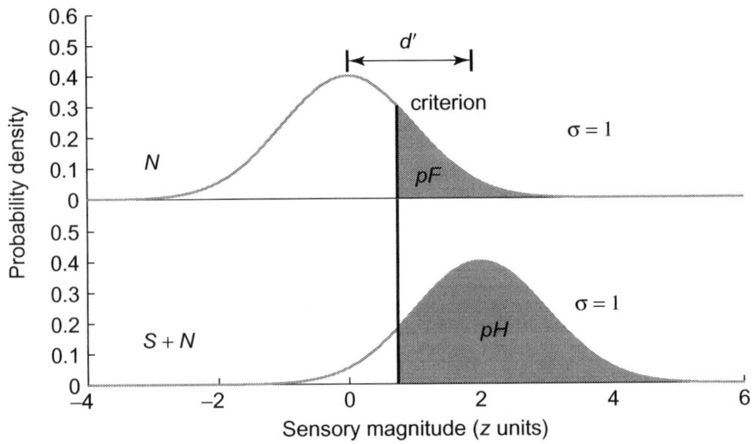

FIGURE 6.6 Distributions of sensory magnitude in response to both noise *N* and signal-plus-noise ($S + N$) in a 1AFC yes/no task. The vertical black line shows the position of the observer's criterion. Sensory magnitudes to the right of this line result in a "yes" response, while those to the left a "no" response. *pH* is the proportion of "hits," or correct "yes" responses, and *pF* the proportion of "false alarms," i.e., incorrect "yes" responses. *pH* and *pF* are given by the gray areas to the right of the criterion line.

trials resulting in a "yes" response, i.e., the proportion of hits or pH. The gray area to the right of the criterion line in the upper N distribution gives the number of "yes" responses in target-absent trials, i.e., the proportion of false alarms or pF. If we denote the position of the criterion line on the abscissa as c (see below), then one can see that:

$$pF = 1 - \Phi(c)$$

$$\text{or} \quad pF = \Phi(-c)$$

and

$$pH = 1 - \Phi(c - d')$$

$$\text{or} \quad pH = \Phi(-c + d')$$

Converting pF and pH to z-values one obtains:

$$z(pF) = -c$$

and

$$z(pH) = -c + d'$$

Combining these two equations and solving for d' gives

$$d' = z(pH) - z(pF) \tag{6.7}$$

6.3.3.2 *Calculation of Criterion C for 1AFC*

In Figure 6.6 it can be seen that the criterion is measurable in z units, with a high z-value implying a strict criterion (few hits but few false alarms), and a low z-value implying a loose criterion (many hits but many false alarms). However, the actual criterion z-value depends on where the zero z-value is positioned, so a convention is needed to ensure that the criterion measure is comparable across conditions. The convention is to place the zero point midway between the N and $S + N$ distributions, as shown in Figure 6.7.

With $z = 0$ centered midway between the two distributions, the criterion, denoted by C, is positioned in the noise distribution at:

$$z(1 - pF) - d'/2$$

and in the signal-plus-noise distribution at:

$$z(1 - pH) + d'/2$$

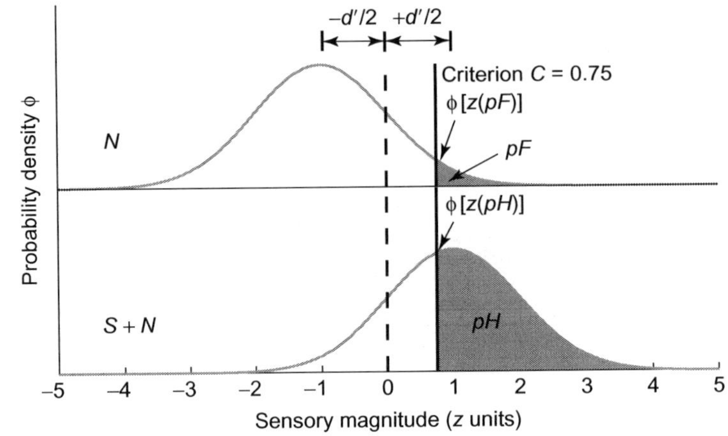

FIGURE 6.7 Method for calculating criterion C. Note that $z = 0$ is centred midway between the N and $S + N$ distributions. See text for further details.

However, since $z(p) = z(1 - p)$, the two expressions can be rewritten as $-z\,(pF) - d'/2$ and $-z\,(pH) + d'/2$. Thus, the position of C can be defined in two ways:

$$C = -z(pH) + d'/2$$

and

$$C = -z(pF) - d'/2$$

Adding the two equations together gives:

$$C = -[z(pH) + z(pF)]/2 \tag{6.8}$$

(Macmillan & Creelman, 2005). Thus, criterion C can be estimated by converting pH and pF into z-values, and then using Equation 6.8. C is calculated this way in **PAL_SDT_1AFC_PHFtoDP**. C can range from negative to positive, with negative values indicating a bias towards "yes" responses and positive values a bias towards "no" responses.

6.3.3.3 Calculation of Criterion lnβ for 1AFC

An alternative measure of the criterion is the natural logarithm of the ratio of the heights of the two distributions at C (Macmillan & Creelman, 2005). The heights at C are shown in Figure 6.7 as $\phi[z(pH)]$ and $\phi[z(pF)]$. Thus:

$$\ln\beta = \ln\frac{\phi[z(pH)]}{\phi[z(pF)]} \tag{6.9}$$

Now $\phi[z(pH)]$ and $\phi[z(pF)]$ are given by:

$$\phi[z(pH)] = \frac{1}{\sqrt{2\pi}}\exp\left[\frac{-\{-z(pH)^2\}}{2}\right] \quad (6.10)$$

and

$$\phi[z(pF)] = \frac{1}{\sqrt{2\pi}}\exp\left[\frac{-\{-z(pF)^2\}}{2}\right] \quad (6.11)$$

Taking natural logarithms of the two equations, i.e., $\ln\{\phi[z(pH)]\}$ and $\ln\{\phi[z(pF)]\}$ and then substituting the results into the equation for $\ln\beta$, simple algebra shows that:

$$\ln\beta = [z(pF)^2 - z(pH)^2]/2 \quad (6.12)$$

This is how $\ln\beta$ is calculated in **PAL_SDT_1AFC_PHFtoDP**. $\text{Ln}\beta$ behaves in the same way as C. The reason for this is that $\ln\beta = Cd'$. Readers can check this relationship themselves using the equations above.

6.3.3.4 *Calculation of Criterion C′ for 1AFC*

A third measure of the criterion not calculated by Palamedes is C', which is C expressed as a proportion of d' (Macmillan & Creelman, 2005):

$$C' = C/d' = \frac{-[z(pH) + z(pF)]}{2[z(pH) - z(pF)]} \quad (6.13)$$

6.3.3.5 *Calculation of Pc_{max} for 1AFC*

In Section A we showed graphically that with a 1AFC task the optimum Pc, or Pc_{max}, is obtained when the observer is unbiased, i.e., when $C = 0$. It follows from Equation 6.8 that when $C = 0$, $z(pH) = -z(pF)$. Since $d' = z(pH) - z(pF)$ (Equation 6.7), simple algebra reveals that when $C = 0$, $d' = 2z(pH)$. Converting $z(pH)$ to Pc_{max} gives:

$$Pc_{\text{max}} = \Phi(d'/2) \quad (6.14)$$

The interested reader may wish to prove that Pc reaches a maximum when $C = 0$. One can determine if the observer is operating optimally in a 1AFC task by testing whether $pH = 1 - pF$. When performing optimally, pH is Pc_{max}, and d' can be calculated as $2z(pH)$.

6.3.4 Calculation of d' for Unbiased and Biased 2AFC Tasks

In the first section of Section B we derived the formula for calculating d' from Pc for an unbiased standard M-AFC task (Equation 6.6). This formula can be used to calculate d' for the standard 2AFC task ($M = 2$), assuming that the observer is unbiased. In the following sections we show a simpler method for calculating d' for an unbiased 2AFC task, and show how d' and measures of bias can be calculated for 2AFC tasks in which the observer is biased.

6.3.4.1 *Alternative Calculation of d' for Unbiased 2AFC*

With the standard 2AFC procedure, the N and $S + N$ stimuli are presented together in a trial as two alternatives. Remember that the decision rule is to choose the alternative in which the internal signal is biggest. If the observer adopts this rule, trials in which the *differences* between the $S + N$ and N samples are positive will result in a correct decision. Now the distribution of differences between random samples from two equal-variance normal distributions, one with mean 0 the other with mean d', is a normal distribution with a mean of d' and a variance of 2, i.e., a σ of $\sqrt{2}$. The σ of $\sqrt{2}$ follows from the variance sum law. This law states that the variance of the sum, or of the difference, between two uncorrelated random variables is the sum of the variances of the two variables. Thus, if the two distributions each have a σ of 1, the σ of the difference between the two distributions is $\sqrt{(1^2 + 1^2)} = \sqrt{2}$. The $(S + N) - N$ *difference* distribution is illustrated in the lower panel of Figure 6.8. Note that in this graph the abscissa is in z units that have been normalized to the σs of the N and $S + N$ distributions, not to the σ of their difference.

The proportion correct for 2AFC is thus given by the gray area in the lower panel to the right of zero. This is:

$$Pc = \Phi\left(\frac{d'}{\sqrt{2}}\right) \tag{6.15}$$

(Wickens, 2002; Macmillan & Creelman, 2005; McNicol, 2002). Equation 6.15 converts a z-value of $d'/\sqrt{2}$ to a Φ value. One can implement the equation using **PAL_ZtoP** using as an input argument a scalar, vector or matrix of $d'/\sqrt{2}$. However, in most instances we want to obtain d' from Pc, so for this we use the inverse equation:

$$d' = z(Pc)\sqrt{2} \tag{6.16}$$

The following converts a vector **PropCorr** containing an array of Pcs into a vector **DP** containing an array of d's, for an unbiased 2AFC task:

```
>>DP=PAL_PtoZ(PropCorr)*sqrt(2)
```

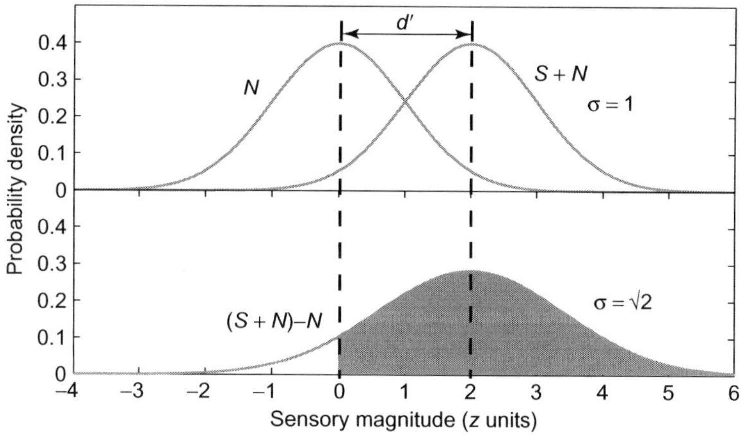

FIGURE 6.8 Graphical illustration of how d' can be calculated for an unbiased 2AFC task. Top: distributions of noise alone (N) and signal-plus-noise ($S + N$) separated by d'. Bottom: distribution of the difference between the two distributions: $(S + N) - N$. Note the different σs for the upper and lower distributions. The z values along the abscissa are normalized to the σ of the two distributions in the top, not bottom panel. See text for further details.

6.3.4.2 *Calculation of d′ for Biased 2AFC*

Let us consider the situation in which the two alternatives are presented sequentially, i.e., 2IFC. Figure 6.9 plots the distribution of differences in sensory magnitude between those in the first interval (X1) and those in the second interval (X2), i.e., the distribution of X1 − X2. Note that there are now two distributions, one for signal present in the first interval and one for signal present in the second interval. The two distributions will be separated by $2d'$ and have σs of $\sqrt{2}$ (see above). If the observer is biased towards responding to one interval more than the other, then their criterion C will be non-zero. The observer's decision rule is "1" (first interval) if X1 − X2 > C, and "2" (second interval) if X1 − X2 < C. As explained in Section A, the key to calculating d' for a biased 2AFC task is to classify the responses in terms of hits and false alarms, where a "1" response is scored as a hit when the signal is in the first interval and a false alarm when the signal is in the second interval.

One can see from Figure 6.9 that:

$$z(pH) = (d' - C)/\sqrt{2} \tag{6.17}$$

and

$$z(pF) = (-d' - C)/\sqrt{2} \tag{6.18}$$

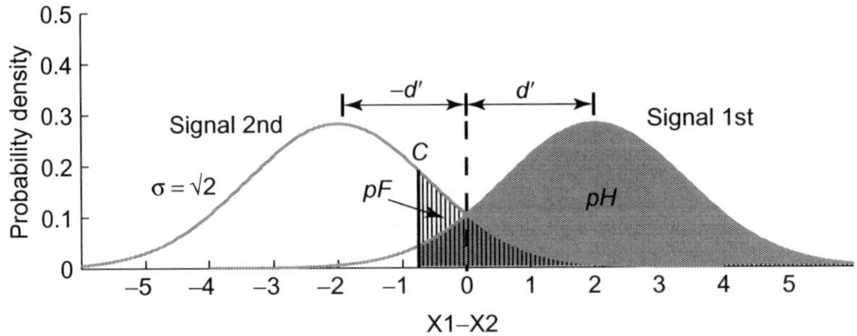

FIGURE 6.9 Relationship between d', C, pH, and pF in a biased 2AFC task. Each plot gives the distribution of differences between the sensory magnitudes in the first (X1) and second (X2) alternatives/intervals. If the signal is in the first alternative/interval the distribution is the one shown on the right, if in the second interval the distribution shown on the left.

Combining the two equations and solving for d' gives:

$$d' = [z(pH) - z(pF)]/\sqrt{2} \tag{6.19}$$

This is the equation used to calculate d' in **PAL_SDT_2AFC_PHFtoDP**.

6.3.4.3 *Calculation of C and lnβ for Biased 2AFC*

Combining Equations 6.17 and 6.18 above and solving for C gives:

$$C = -[z(pH) + z(pF)]/\sqrt{2} \tag{6.20}$$

The criterion measure lnβ is defined in Equation 6.9, and it is important to note that the ϕs refer to the heights of $z(pH)$ and $z(pF)$ in the standard normal distribution, i.e., the distributions in the upper panel of Figure 6.8, not to the heights of the difference distributions in Figure 6.9. The calculation of lnβ for the 2AFC task is thus identical to that for the 1AFC task (we would like to thank Mark Georgeson for pointing this out), and is hence given by Equation 6.12. Equation 6.20 is used to calculate bias for the standard 2AFC task by **PAL_SDT_2AFC_PHFtoDP**.

6.3.4.4 *Calculation of Pc_{max} for 2AFC*

From Equation 6.20, if $C = 0$, then $z(pH) = -z(pF)$. Combining this result with Equation 6.19 reveals that when $C = 0$, $d'\sqrt{2} = 2z(pH)$. Converting $z(pH)$ to $Pc_{\max}$ gives:

$$Pc_{\max} = \Phi(d'/\sqrt{2}) \tag{6.21}$$

6.3.5 Calculation of d' for Same-Different Tasks

For the calculation of d' for a same-different task we adopt the convention of referring to the two relevant distributions as S_1 and S_2 (signal 1 and 2), rather than N and $S + N$. It would be unusual to employ a same-different task to measure the detectability of a target when the alternative location/interval was a blank. The same-different task is most appropriate to situations in which the observer is required to discriminate two suprathreshold stimuli without necessarily having to know the basis of the discrimination.

6.3.5.1 *Calculation of d' for a 2AFC Same-Different*

The computation of d' for the same-different task in which both the same *and* different pairs are presented together during a trial is described by Macmillan, Kaplan, and Creelman (1977). They use the term 4IAX to characterize the task, since they consider the scenario in which the four stimuli are presented in temporal order, as in an auditory experiment.

Let us begin with the standard assumption that the sensory magnitudes of S_1 and S_2 are normally distributed and separated by d'. According to Macmillan et al. (1977), the most likely strategy employed by observers in this task is to compare the absolute difference between the two signals in each of the first and second pairs. The observer responds "1" if the difference between the first pair is perceived to be greater than the difference between the second pair, and "2" otherwise. Suppose that the sensory magnitudes of the four stimuli are represented by the sequence X1, X2, X3, and X4. The decision rule is therefore to respond "1" if $|X1 - X2| > |X3 - X4|$, and "2" if $|X1 - X2| < |X3 - X4|$.

Figure 6.10, adapted from Macmillan et al. (1977), illustrates the computation of d' for the task. The abscissa and ordinate in Figure 6.9 represent, respectively, the decision variables $X1 - X2$ and $X3 - X4$. The gray areas in the figure represent the combinations of decision variables that result in a "1" decision, i.e., areas where $|X1 - X2| > |X3 - X4|$. The gray areas can be subdivided into four regions: upper left; lower left; upper right; and lower right. On the right side of the figure the gray area defines the space in which $X1 - X2$ is more positive than either $X3 - X4$ (upper right) or $-(X3 - X4)$ (lower right). On the left of the figure the gray area defines the space in which $X1 - X2$ is more negative than either $X3 - X4$ (lower left) or $-(X3 - X4)$ (upper left).

The observer will be correct when making a "1" decision if the samples that fall within the gray regions are from any of the following sequences: $<S_1S_2S_1S_1>$, $<S_1S_2S_2S_2>$, $<S_2S_1S_1S_1>$ or $<S_2S_1S_2S_2>$. On the other hand, the observer will be incorrect when responding "1" if samples from the remaining sequences fall within the gray area, namely $<S_1S_1S_1S_2>$, $<S_1S_1S_2S_1>$, $<S_2S_2S_1S_2>$ or $<S_2S_2S_2S_1>$. *Pc* is therefore the probability that samples from the first four sequences will fall within either of the two (left or right) gray areas. The four rings in the figure denote

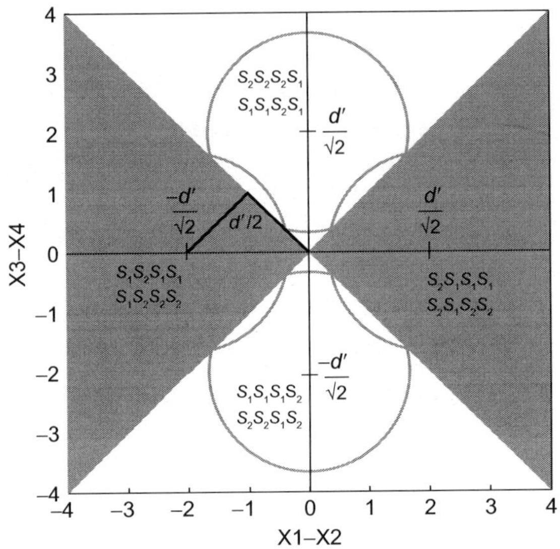

FIGURE 6.10 Graphical representation of the distributions involved in the 2AFC same-different task. X1 . . . X4 represent the internal sensory magnitudes of the four stimuli. Note that the abscissa plots X1 − X2 and the ordinate X3 − X4. The sequences <$S_1S_2S_1S_1$> etc., denote joint sample distributions of stimulus sequences. Note that the distance to the center of each distribution from the center of the figure is $d'/\sqrt{2}$, but when measured from a point on the diagonal perpendicular to the center of each distribution (shown by the thick black lines in the upper left quadrant) the distance is $d'/2$. The figure is adapted from Figure 6a in Macmillan et al., (1977).

volumes of the joint likelihood distributions of the various sequences of S_1 and S_2. Note that the σ of the distributions is $\sqrt{2}$, because they are distributions of the difference between samples from two normal distributions.

Each volume in the left and right gray areas comprises two probabilities, a "small" and a "large." The large probability is the probability that samples from the sequences specified within each gray area of the figure will fall within that area. However, there is a small probability that samples from the sequences in the opposite gray area will also fall within the area. For example, although most of the samples that fall within the gray area on the right of the figure will come from sequences <$S_2S_1S_1S_1$> and <$S_2S_1S_2S_2$>, a few will come from <S_1,S_2,S_1,S_1> and <S_1,S_2,S_2,S_2>. This is because even though most of the difference signals $S_1 - S_2$ are "large negative" and hence fall within the gray area on the left, a few will be "large positive" and will fall within the gray area on the right. Remember that it does not matter whether the difference $S_1 - S_2$ is "large negative" or "large positive," as long as its absolute magnitude is greater than $S_2 - S_2$ or $S_1 - S_1$ (the possible sequences in the other alternative/interval). Either way the response "1" will be correct.

The larger probability within each gray area is given by $[\Phi d'/2]^2$ while the smaller probability is given by $[\Phi(-d'/2)]^2$. The denominator of 2 in each expression reflects the fact that the area described by the gray rectangle has sides that, by the Pythagorean Theorem, extend by $d'/2$ to the midpoint of the distribution along the side, as illustrated in the upper left quadrant of the figure. The squaring of each expression reflects the fact that one is dealing with a bivariate, i.e., joint, distribution. To obtain Pc we simply add together the large and small probabilities:

$$Pc = [\Phi(d'/2)]^2 + [\Phi(-d'/2)]^2 \tag{6.22}$$

Equation 6.22 is used to calculate Pc in **PAL_SDT_2AFCsameDiff_DPtoPC**. Following Macmillan and Creelman (2005), the equation can be inverted to obtain d' from Pc using:

$$d' = 2z[0.5\{1 + (2Pc - 1)^2\}] \tag{6.23}$$

and this is used to calculate d' in **PAL_SDT_2AFCsameDiff_PCtoDP**.

6.3.5.2 *Calculation of d′ for a 1AFC Same-Different Task: Differencing Model*

In the differencing model of the 1AFC same-different task it is assumed that the observer encodes the perceived difference between the two stimuli in the trial, and if the absolute value of the difference exceeds a criterion the observer responds "different," if not "same." Suppose the signal from the first stimulus is X1, and from the second stimulus X2. The decision rule is therefore "different" if $|X1 - X2| > k$, where k = the criterion, and "same" otherwise. As with the 2AFC same-different task discussed in the previous section, it is useful to consider both the positive and negative parts of the difference signal X1 − X2. The top of Figure 6.11 shows the distributions of sensory magnitudes for the two stimuli S_1 and S_2, centered on 0. The middle and bottom panels (adapted from Figure 9.5 in Macmillan & Creelman, 2005) show the relative likelihoods of the various stimulus pairs as a function of the decision variable X1 − X2. The middle panel shows the Same distributions $<S_1S_1>$ and $<S_2S_2>$, and the bottom panel the Different distributions $<S_1S_2>$ and $<S_2S_1>$.

All the Different distributions have a σ of $\sqrt{2}$, in accordance with the variance sum law. To understand how pH and pF are calculated the criterion has been placed to one side of the midpoint. Given the particular value of d' and k in the figure, most of the $<S_2,S_1>$ signals fall above the criterion k and constitute a "large" probability. Although most of the $<S_1S_2>$ signals fall to the left of k, a few will be "large positive" and fall to its right. As with the 2AFC same-different task we have to include the small probability in the calculation, because it accords with the adopted decision rule. From Figure 6.11 the proportion of hits, pH is given by:

$$pH = \Phi[(d' - k)/\sqrt{2}] + \Phi[(-d' - k)/\sqrt{2}] \tag{6.24}$$

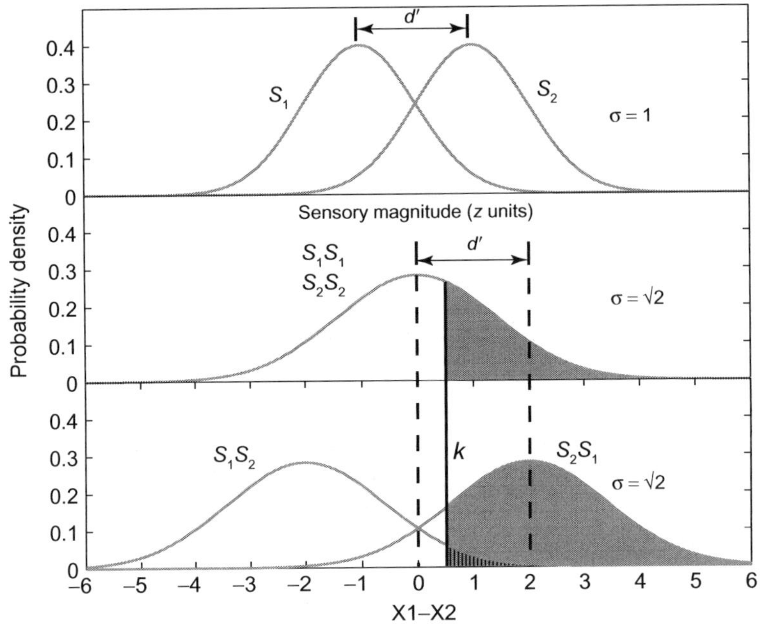

FIGURE 6.11 Method for calculating d' for a 1AFC same-different task assuming a differencing model. See text for details.

where the larger of the two terms is given by the gray area to the right of k and the smaller of the two terms by the hatched area to the right of k. The proportion of false alarms pF is given by the area to the right of the criterion line in the middle panel, multiplied by 2 since there are two distributions, i.e.:

$$pF = 2\Phi(-k/\sqrt{2}) \tag{6.25}$$

The routine **PAL_SDT_1AFCsameDiff_DiffMod_DPtoPHF** performs these calculations. To calculate d' and k from pH and pF, as is mostly required, the routine **PAL_SDT_1AFCsameDiff_DiffMod_PHFtoDP** exploits the fact that k can be obtained directly from pF, as from Equation 6.24 $k = -z(pF/2)\sqrt{2}$. The value of k is then substituted into Equation 6.24 and the routine performs an iterative search to find that value of d' that results in the input value of pH. Further details of the 1AFC same-different differencing model can be found in Macmillan and Creelman (2005).

6.3.5.3 *Calculation of d′ for a 1AFC Same-Different Task: Independent Observation Model*

According to Macmillan and Creelman (2005), the observer's optimum strategy for the same-different task in which only two stimuli are presented per trial

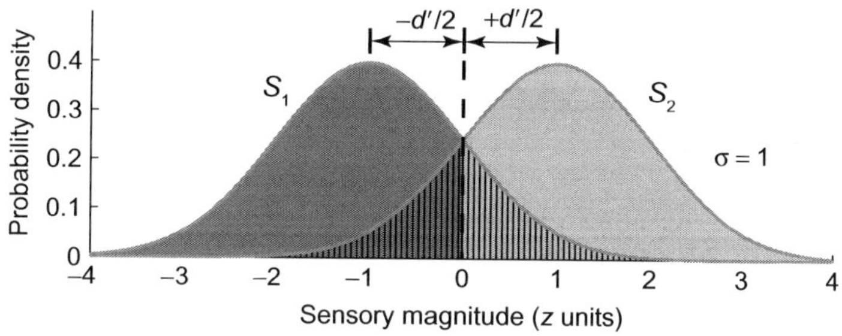

FIGURE 6.12 Principle behind the computation of d' for the independent observation model of the 1AFC same-different task. See text for details.

is to respond "different" when the signals from S_1 and S_2 fall on either side of a criterion centered midway between the two distributions. They term this model the independent observation model. To compute d' for this model, Macmillan and Creelman suggest the following method. First, calculate the Pc that an observer would obtain for this task if they were operating optimally (this is Pc_{max}) using pH and pF. Second, use Pc_{max} to compute d'. Third, use the values of pH and pF to compute the criterion C in the same way as for the standard 1AFC task.

As elsewhere, it is best to begin with the method for calculating Pc_{max} from d', rather than the reverse. Macmillan and Creelman (2005) provide a three-dimensional representation of the decision space for the independent observation model, as the calculations involve joint likelihood distributions. The two-dimensional representation provided in Figure 6.12 should, however, be sufficient to understand the principle behind the calculation.

In Figure 6.12, the probability that signals from both S_1 and S_2 will fall on opposite sides of the criterion at zero is the probability that S_1 falls to the left of the criterion multiplied by the probability that S_2 falls to its right (since we are dealing here with the joint probability of an event). In the figure, given the value of $d' = 2$, most of the S_2 signals fall to the right of the criterion and most of the S_1 signals will fall to the left of the criterion, so the product of the two signals will be a "large" probability given by $[\Phi(d'/2)]^2$. However, there is a small probability that both a high value of S_1 *and* a low value of S_2 will fall on either side of the criterion. These probabilities are the smaller hatched areas in the figure. The observer will also be correct in these instances, since the decision rule is to respond "different" when the signals from the two stimuli fall on either side of the criterion. The joint probability in this case is given by the product of the hatched areas, which is $[\Phi(-d'/2]^2$. Thus, to obtain Pc_{max} we add up the two joint probabilities:

$$Pc_{max} = [\Phi(d'/2)]^2 + [\Phi(-d'/2)]^2 \tag{6.26}$$

and from this equation, d' is given by:

$$d' = 2z\{0.5[1 + \sqrt{2Pc_{max} - 1}]\} \tag{6.27}$$

(Macmillan & Creelman, 2005). To calculate d' from pH and pF, Pc_{max} is first estimated using:

$$Pc_{max} = \Phi\{[z(pH) - z(pF)]/2\} \tag{6.28}$$

and the result substituted into Equation 6.27. This calculation is performed by **PAL_SDT_1AFCsameDiff_IndMod_PHFtoDP**. The same routine also calculates the observer's criterion using $C = -0.5[z(pH) + z(pF)]$. The reverse calculation (pH and pF from d' and C) is performed by **PAL_SDT_1AFCsameDiff_IndMod_DPtoPHF**.

6.3.6 Calculation of d′ for Match-to-Sample Tasks

6.3.6.1 Calculation of d′ for 2AFC Match-to-Sample: Independent Observation Model

The computation of d' for the 2AFC match-to-sample task under the independent observation model parallels that of the 1AFC same-different task. According to Macmillan and Creelman (2005), who refer to the task as ABX, Pc for an unbiased observer is given by:

$$Pc = \Phi(d'/\sqrt{2})\cdot\Phi(d'/2) + \Phi(-d'/\sqrt{2})\cdot\Phi(-d'/2) \tag{6.29}$$

We refer readers to Macmillan and Creelman (2005) for the derivation of this equation. The calculation is performed by **PAL_SDT_2AFCmatchSample_IndMod_DPtoPC**. The inverse routine **PAL_SDT_2AFCmatchSample_IndMod_DPtoPC** employs an iterative search procedure using Equation 6.29 to obtain d' from Pc. If the raw data are hits and false alarms, defined according to the rule for a conventional 2AFC task, **PAL_SDT_2AFCmatchSample_IndMod_PHFtoDP** first calculates Pc_{max} and then d' by iterative search of Equation 6.29. **PAL_SDT_2AFCmatchSample_IndMod_DPtoPHF** performs the reverse calculations.

6.3.6.2 Calculation of d′ for 2AFC Match-to-Sample: Differencing Model

For the 2AFC match-to-sample differencing model, the observer is assumed to encode the difference in sensory magnitude between the Sample and each of the Match stimuli, and choose the Match with the smallest absolute Sample-minus-Match difference. According to Macmillan and Creelman (2005) the differencing strategy, as with the same-different task, is the more likely to be adopted in a roving experiment where many different stimuli are presented during a session. Macmillan and Creelman (2005) have derived the following equation for the unbiased observer:

$$Pc = \Phi(d'/\sqrt{2}) \cdot \Phi(d'/\sqrt{6}) + \Phi(-d'/\sqrt{2}) \cdot \Phi(-d'/\sqrt{6}) \quad (6.30)$$

Palamedes performs the calculation in **PAL_SDT_2AFCmatchSample_DiffMod_DPtoPC**. The inverse routine **PAL_SDT_2AFCmatchSample_DiffMod_PCtoDP** calculates d' by iterative search of Equation 6.30. If the data are hits and false alarms, d' and criterion C can be obtained using **PAL_SDT_2AFCmatchSample_DiffMod_PHFtoDP**, whose inverse is **PAL_SDT_2AFCmatchSample_DiffMod_DPtoPHF**. The calculations in these routines parallel those for the 2AFC match-to-sample independent observation model.

6.3.6.3 *Calculation of d′ for M-AFC Match-to-Sample*

For match-to-sample tasks in which $M > 2$, the Palamedes routines **PAL_SDT_MAFCmatchSample_DPtoPC** and **PAL_SDT_MAFCmatchSample_PCtoDP** assume an unbiased observer and a differencing strategy. In the first routine, Pc is computed by Monte Carlo simulation rather than by equation. The simulation works as follows. Let S_1 and S_2 represent the two signal distributions separated by d'. Let S_1 be the Sample stimulus. Therefore, the M Match stimuli comprise one S_1 and $M - 1$ S_2s. On each "trial" of the simulation, two random samples (don't confuse a random "sample" with the "Sample" stimulus!) are selected from S_1 (one for the Sample stimulus and one for the Match stimulus), and $M - 1$ random samples are selected from S_2 (the other Match stimuli). The absolute difference between the Sample S_1 and each of the M Match stimuli is then calculated. If the absolute difference between the Sample S_1 and the Match S_1 is smaller than all of the absolute differences between the Sample S_1 and Match S_2s, then by the differencing strategy the trial is scored "correct," otherwise "incorrect." The process is then repeated over a large number of trials and the overall proportion correct calculated across trials. Note that if the sample stimulus was designated to be S_2 the result would be expected to be the same, so the choice of S_1 or S_2 as the Sample is arbitrary. Because the routine uses Monte Carlo simulation, the computed Pc will not be identical each time, but should be accurate to about two decimal places. The routine is also relatively slow owing to the large number of trial simulations involved. The inverse routine that calculates d' from Pc is especially slow since it involves an iterative search of the forward routine.

6.3.7 Calculation of *d′* for M-AFC Oddity Tasks

The Palamedes routine **PAL_SDT_MAFCoddity_DPtoPC** also assumes an unbiased observer and a differencing strategy. It calculates Pc by Monte Carlo simulation following the method described by Craven (1992). Let S_1 and S_2 represent the two signal distributions separated by d'. Assume that S_1 is the oddity. Therefore, there are $M - 1$ non-oddity S_2s. On each trial of the simulation, a random sample is selected from the S_1 oddity and each of the $M - 1$ non-oddity S_2s. Consider that every random sample is a possible oddity. Recall that the decision rule is to select the alternative most different from the average of all the alternatives. We therefore calculate the absolute difference between each sample and the average of all the M samples. According to the decision rule, the sample selected to be the oddity is the one with the biggest absolute difference. If this sample is from S_1 then the trial is scored "correct," else "incorrect." The process is repeated over a large number of trials and the proportion correct calculated across trial. As with the routines for M-AFC match-to-sample, the computed Pc will not be identical each time, but should be accurate to about two decimal places. The routines are also relatively slow owing to the large number of trial simulations involved.

Further Reading

The best starting points for SDT are McNicol (2004), Chapters 5–8 of Gescheider (1997) and Macmillan & Creelman (2005). The most comprehensive treatment of SDT that is accessible to the non-expert, is Macmillan & Creelman (2005). More mathematical treatments can be found in Wickens (2002) and Green & Swets (1974). Further details of the computation of d' for the same-different tasks can be found in Macmillan et al. (1977).

Exercises

1. Consider the $M > 2$ versions of the standard forced-choice, oddity and match-to-sample tasks. The Palamedes routines for the M-AFC versions of these tasks assume that there are just two stimuli, S_1 and S_2, and that the observer is unbiased and employs the following decision rules: for the standard forced-choice task select the alternative with the largest stimulus magnitude; for the oddity task select the alternative most different from the mean of all the alternatives; for the match-to-sample task select the match most similar to the sample. For a given d', which task would you expect to produce the biggest and which the smallest Pc? Write a script using the Palamedes routines to plot Pc against M for a given d' for each task to test your predictions.
2. The following exercise emerged from discussions with Mark Georgeson. Table 6.2 presents the results of an experiment aimed at measuring a psychometric function of proportion correct against stimulus magnitude using a standard 2AFC task. The experimenter is interested in the effects of bias on the estimates of the threshold and slope of the psychometric function, so the results are presented in terms of proportion hits pH and proportion false alarms pF, as calculated according to the rules in Section 6.2.4.

TABLE 6.2 Results of a hypothetical experiment aimed at deriving a psychometric function using a standard 2AFC task

Stimulus magnitude	*pH*	*pF*
1	0.61	0.53
2	0.69	0.42
3	0.79	0.33
4	0.88	0.18
5	0.97	0.06
6	0.99	0.03

Use the appropriate Palamedes routines to calculate d', criterion C, and proportion correct Pc, for each pair of pH and pF. Then calculate the Pc_{max} for each stimulus magnitude that would be expected if the observer was unbiased (see Section 6.2.8). Plot psychometric functions of both Pc and Pc_{max} against stimulus magnitude (Chapter 4) and obtain estimates of the thresholds and slopes of the functions. Are the thresholds and slopes significantly different for the two functions (see Chapter 8)?

References

Craven, B. J. (1992). A table of d' for M-alternative odd-man-out forced-choice procedures. *Perception & Psychophysics*, *51*, 379–385.

Elliot, P. B. (1964). Tables of d'. In J. A. Swets (Ed.), *Signal Detection and Recognition by Human Observers*. New York: Wiley.

Gescheider, G. A. (1997). Psychophysics: The Fundamentals.. Mahwah, New Jersey: Lawrence Erlbaum Associates.

Green, D. A., & Swets, J. A. (1974). *Signal Detection Theory and Psychophysics*. Huntington, New York: Krieger.

Kingdom, F., Moulden, B., & Hall, R. (1987). Model for the detection of line signals in visual noise. *Journal of Optical Society of American A-Optics Image Science and Vision*, *4*, 2342–2354.

Macmillan, N. A., & Creelman, C. D. (2005). Detection Theory: A User's Guide. Mahwah, New Jersey: Lawrence Erlbaum Associates.

Macmillan, N. A., Kaplan, H. L., & Creelman, C. D. (1977). The psychophysics of categorical perception. *Psychological Review*, *84*, 452–471.

McNicol, D. (2004). A Primer of Signal Detection Theory. Mahwah, New Jersey: Lawrence Erlbaum Associates.

Wickens, T. D. (2002). Elementary Signal Detection Theory. Oxford, New York: Oxford University Press.

Yeshurun, Y., Carrasco, M., & Maloney, L. T. (2008). Bias and sensitivity in two-interval forced procedures. *Vision Research*, *48*, 1837–1851.

CHAPTER

7

Scaling Methods

7.1 INTRODUCTION

Perceptual scales, sometimes termed "psychological scales," "sensory scales," or "transducer functions," describe the relationship between the perceived and physical magnitudes of a stimulus. Example perceptual scales are the relationships between perceived contrast and physical contrast, perceived depth and retinal disparity, perceived velocity and physical velocity, and perceived transparency and physical transparency. In Chapter 3, Section 3.3.2 summarizes many of the methods available for measuring perceptual scales, and we recommend that this section be read before the rest of this chapter.

Perceptual scales are in most cases descriptions of stimulus appearance, and are thus derived from procedures that have no correct and incorrect answer on each trial, in other words Type 2 according to the taxonomy outlined in Chapter 2. For certain scaling tasks this might seem counterintuitive. Take the method of paired comparisons, in which the observer is required on each trial to decide which of two stimuli appears greater in perceived magnitude along the dimension of interest. If the growth of perceived magnitude is a monotonically increasing function of stimulus magnitude, one can legitimately argue that the observer's judgement is "correct" when the chosen stimulus is the one with the higher physical magnitude. The argument does not hold, however, for scales which are not monotonic, as for example in the color wheel where the colors are arranged around a circle. Moreover, for perceptual scaling methods involving comparisons of stimuli from widely different parts of the stimulus range, such as the method of triads or method of quadruples, it is meaningless to consider the observer's responses in terms of correct/incorrect unless the scale is perfectly linear, which in most cases it is not. That being said, not all perceptual scales are derived from appearance-based judgements. Fechnerian or Discrimination scales, of which more will be said in Section B, are derived by integrating JNDs (specifically increment thresholds) across the stimulus range, and are therefore performance-based.

To illustrate the general principle of a perceptual scale consider Figure 7.1. In this hypothetical example perceived stimulus magnitude, denoted by ψ, is described by a power function whose general form is aS^n, where S is the stimulus level, a an arbitrary scaling factor, and n an exponent. In the figure $n = 0.5$. If $n < 1$ the exponent determines how bow-shaped, or "compressive," the function is, whereas if $n > 1$ the exponent determines how expansive it is. In the majority of scaling methods, observers are required to make judgements about combinations of stimuli selected from various parts of the stimulus range. For example, in the method of quadruples, observers compare two pairs of stimuli on each trial and decide which pair appears

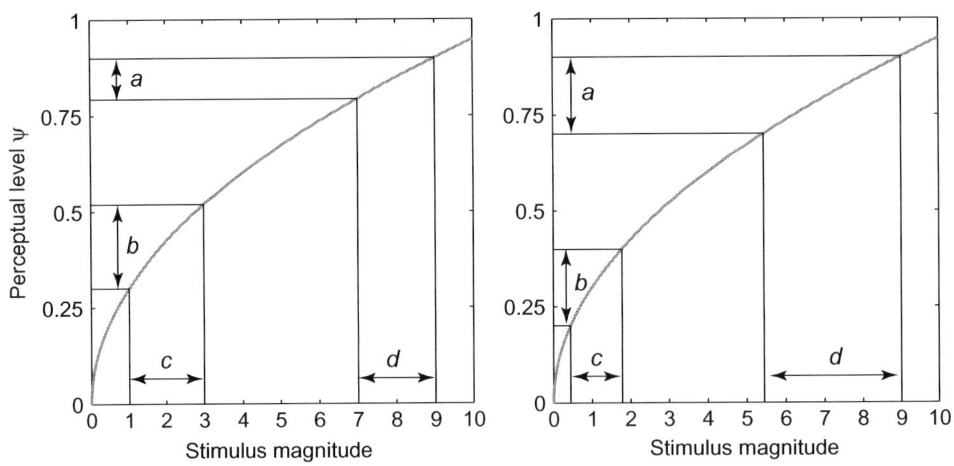

FIGURE 7.1 Hypothetical perceptual scale. Left: two pairs of stimuli with the same physical difference ($c = d$) produce different values of perceived difference ($b > a$). Right: two pairs of stimuli with different physical difference ($d > c$) are equally different perceptually ($a = b$).

more similar (or more different). Figure 7.1a illustrates how two pairs of stimuli, with magnitudes 1 and 3, and 7 and 9, will differ in their perceived similarity (or difference) owing to the compressive nature of the scale, with the 7 and 9 pair appearing more similar than the 1 and 3 pair. Conversely, Figure 7.1b shows how two pairs of stimuli that differ in the physical magnitude of their differences nevertheless can appear equally similar in terms of the perceptual magnitude of their differences.

In this chapter we describe in some detail a scaling method termed Maximum Likelihood Difference Scaling, or MLDS (Maloney & Yang, 2003). MLDS is a relatively new method and has some very attractive features. It avails itself to forced-choice scaling methods and exploits state-of-the-art computer optimization algorithms for parameter estimation. It also appears to be robust to whether the internal noise of the observer associated with each stimulus magnitude is a constant, or grows with stimulus magnitude. This last property of MLDS is an important one, but its significance must wait until Section B. In promoting MLDS we do not wish to argue that it is the only valid perceptual scaling method. The partition scaling methods described in Chapter 3 offer some advantages over MLDS, and we shall discuss these again in Section B.

MLDS produces "interval" perceptual scales. Remember from Chapter 3 that with an interval scale it is the differences between scale values rather than the values themselves that characterize the underlying perceptual representation. For example, stimulus velocities of 2, 4, and 6 degrees might be represented on an interval perceptual scale by values 1, 5, and 6. This would capture the observation that the perceived difference between 2 and 4 degrees, a difference of 4 units on the perceptual scale, is four times greater than the perceived difference between the 4 and 6 degrees, a difference of 1 unit on the perceptual scale. The same velocities could just as well however be represented by scale values of 4, 12, and 14, as these embody the same difference-relations as 1, 5, and 6. As we noted in Chapter 3, an interval scale can be transformed without loss of information by the equation $aX + b$, where X is the scale value, and a and b are constants. In Figure 7.1 the perceptual scale ranges from 0–1, but could be rescaled to range from 0–100, or 1–10, or any other range for that matter.

As with the other data analysis chapters in this book, the remainder of the chapter is divided into two sections. Section A describes the Palamedes routines used to derive perceptual scales using MLDS. Section B describes the theory behind MLDS as well as partition scaling methods, and explores their underlying assumptions, strengths, and limitations.

7.2 SECTION A: PRACTICE

7.2.1 Maximum Likelihood Difference Scaling (MLDS)

The Palamedes routines for deriving a perceptual scale use the method developed by Maloney and Yang (2003) termed Maximum Likelihood Difference Scaling, or MLDS. Although Maloney and Yang (2003) used MLDS in conjunction with the method of quadruples, we have extended it for use with paired comparisons and the method of triads.

Let us first remind ourselves of the observer's task in each of these three methods (and see Figure 3.3). With paired comparisons the observer is presented on each trial with *two* stimuli, say A and B, drawn from a larger set, and decides which stimulus is greater in perceived magnitude or "further along" the dimension of interest. With the method of triads the observer is presented on each trial with *three* stimuli, say A, B, and C, and decides whether the perceived difference between A and B is larger (or smaller) than the perceived difference between B and C. With the method of quadruples the observer is presented on each trial with *four* stimuli, say A, B, C, and D, and decides whether the perceived difference between A and B is greater (or smaller) than the perceived difference between C and D. All three methods are thus two-alternative forced-choice (2AFC), but remember that because these are appearance-based tasks there is no correct or incorrect answer on any trial.

In order to use MLDS the stimulus space must be sampled in such a way that for any particular stimulus combination not every trial produces the same response. In other words, for any given stimulus combination, we want the observer to choose one of the two alternatives over the other only a *proportion* of times. If the stimulus space is sampled in such a way that all responses for a given stimulus combination are identical, MLDS will fail and only an ordinal scale can be derived. Since MLDS derives perceptual scales from proportions of response judgements, its application to paired comparison data has a formal similarity to Thurstone's (1927) classic method of deriving perceptual scales from paired comparisons (see also Gescheider, 1997).

The best way to understand how to use the Palamedes MLDS routines is to demonstrate their operation using simulated data sets. The simulated data we describe below consists of responses that a hypothetical observer would be expected to make *if* their judgements were determined by an underlying perceptual scale of a particular specified shape. We can then see how MLDS, which makes no assumptions at all about the shape of the underlying perceptual scale, reconstructs the scale from the data. So we begin with the Palamedes routines that *generate hypothetical data*.

7.2.1.1 Generating Stimulus Sets for MLDS

The first step is to generate the stimulus set, and the routine that does this is **PAL_MLDS_GenerateStimList**. Note that this routine is useful not only for helping to demonstrate MLDS, but also for generating stimulus lists for use in actual scaling experiments. The routine is executed as follows:

```
>>StimList = PAL_MLDS_GenerateStimList(N, NumLevels,…
MaxDiffBetweenLevels, NumRepeats);
```

The argument **N** defines the number of stimuli per trial, and should be set to 2, 3 or 4 depending on whether one wishes to generate pairs, triads or quadruples. **NumLevels** is the number of different stimulus magnitudes, or levels.

MaxDiffBetweenLevels is a very useful parameter that precludes the generation of stimulus combinations that are "too far apart," and that would tend to result in identical observer responses across trials. The precise meaning of the parameter depends on whether one is dealing with pairs, triads or quadruples. With pairs, setting **MaxDiffBetweenLevels** to 3 precludes stimulus pairs that are different by more than 3 stimulus levels. So, for example, if there are 10 stimulus levels, the pairs 6 and 9, 1 and 2, and 8 and 10 will appear in the list, but the pairs 2 and 7, 5 and 9, and 3 and 10 will not. With triads, **MaxDiffBetweenLevels** sets an upper limit for the difference-between-the-difference-between stimulus levels. For example if **MaxDiffBetweenLevels** is again set to 3, the triad 1, 6, and 9 would be allowed since $|6 - 1| - |9 - 6| < 3$; the triad 2, 7, and 9 would be allowed since $|7 - 2| - |9 - 7| = 3$; but the triad 2, 8, and 9 would be precluded since $|8 - 2| - |9 - 8| > 3$. The principle for quadruples is the same as for triads. Finally, the argument **NumRepeats** sets the number of repeat trials for each pair/triad/quadruple. The list of pairs/triads/quadruples generated by the routine is stored in the output matrix **StimList**. As an example, type and execute the following:

```
>> StimList = PAL_MLDS_GenerateStimList(3,6,3,1);
```

Now type and execute:

```
>> StimList
```

and the output should be:

```
StimList =
1 2 3
1 2 4
1 2 5
1 2 6
1 3 4
1 3 5
1 3 6
1 4 5
1 4 6
1 5 6
2 3 4
2 3 5
2 3 6
2 4 5
2 4 6
2 5 6
3 4 5
3 4 6
3 5 6
4 5 6
```

Confirm for yourself that the combinations listed are permissible given the value you used for **MaxDiffBetweenLevels**. It must be remembered that **StimList** only lists the stimulus combinations that are to be used in the experiment. The *order* in which they are presented to the observer must of course be randomized, as also must be the order of presentation of the stimuli in each combination.

How many pairs/triads/quadruples will be generated? If all possible combinations are allowed (to achieve this one simply sets **MaxDiffBetweenLevels** to **> =NumLevels-1**), and each combination is listed in only one order (as in the routine here), the binomial coefficient provides the answer. If the number of stimulus levels is S and the number of stimuli per combination N, the total number of unique combinations T is given by:

$$T = \frac{S!}{N!(S-N)!} \tag{7.1}$$

(remember $S! = S\times(S-1)\times(S-2)\times \ldots 1$). Thus, in the present example where $S = 10$ and $N = 2$, T is 45. With pairs ($N = 2$) a simpler formula that gives the same result is $(S^2 - S)/2$.

Try generating other stimulus sets for $N = 2$, 3, and 4. Check the number of combinations generated (use **>>length(StimList)**) against the number calculated using the above equation (don't forget to take into account **NumRepeats** if set to greater than 1). Then try varying **MaxDiffBetweenLevels** and observe the effect on the number of stimulus combinations.

For the next step in our hypothetical experiment, we will again use triads as an example. Execute **PAL_MLDS_GenerateStimList** with arguments 3, 10, 3, and 30. Use **>>StimList** to type out the list of stimulus pairs. Note that each pair is repeated 30 times. This should enable a sufficient number of responses to be simulated for the MLDS fitting routine.

7.2.1.2 *Simulating Observer Responses for MLDS*

Having generated the stimulus list, the next step is to simulate the hypothetical observer's responses. Let us suppose that the underlying shape of the perceptual scale is a Logistic function, which we have seen in Chapter 4 has a sigmoidal shape. First we need to set up the hypothetical perceptual scale values that will determine the simulated responses. Type the following command:

```
>>PsiValuesGen = PAL_Logistic([5 1 0 0],[1:10]);
```

The first argument is a vector of four parameters that defines the shape of the Logistic function (see Chapter 4), and the second argument is a vector defining the stimulus levels. The output **PsiValuesGen** is a vector containing the hypothetical perceptual scale values that correspond to each stimulus level, given the perceptual scale's logistic shape. Next, we need to define a vector **OutOfNum** that lists, for each of the entries in **StimList**, how many trials are to be simulated. Since each entry in **StimList** corresponds to 1 trial, we fill it with 1s.

```
>>OutOfNum = ones(1,size(StimList,1));
```

We can now generate hypothetical responses using **PAL_MLDS_SimulateObserver**. Every response is either "0" or "1," according to the following rules. For pairs, the response is "1" if the first member of each pair is perceived (hypothetically) to be of greater magnitude, otherwise the response is "0." For triads and quadruples, the response is "1" if the first pair is perceived to be more different than the second pair (or the second pair more similar than the first pair), and "0" otherwise. Execute the routine by typing:

```
>>Response = PAL_MLDS_SimulateObserver(StimList, OutOfNum,...
PsiValuesGen, 0.3);
```

The last argument specifies the hypothetical noise level of the decision process and for the present example can be set to 0.3. This is essential. If there were no internal decision noise, our hypothetical observer's responses would be "0" on every trial and MLDS could not be used.

Next we need to combine responses across repeat trials using the **PAL_MLDS_GroupTrialsbyX** routine. Type and execute:

```
>>[StimList NumPos OutOfNum] = PAL_MLDS_...
GroupTrialsbyX(StimList, Response, OutOfNum);
```

The summed responses are contained in the output parameter **NumPos**. The output parameter **OutOfNum** gives the number of trials for each stimulus combination. You might like to look at the results. To view the summed responses for each pair type and execute the following:

```
>>Results = [StimList(:,1),StimList(:,2),...
StimList(:,3),NumPos',OutOfNum']
```

If you had generated pairs you would need to type:

```
>>Results = [StimList(:,1),StimList(:,2), NumPos',OutOfNum']
```

and if quadruples:

```
>>Results = [StimList(:,1),StimList(:,2),...
StimList(:,3),StimList(:,4),NumPos',OutOfNum']
```

Don't forget the inverted commas after **NumPos** and **OutOfNum**, as these are needed to transpose the vectors from rows into columns. Having simulated our experiment, we can now proceed to fitting the data using MLDS.

7.2.1.3 Fitting the Data with MLDS

An important feature of MLDS is that it makes no assumptions as to the shape of the underlying perceptual scale. The parameters fitted by MLDS are not parameters of a pre-defined function shape, as when fitting a psychometric function (see Chapter 4). Instead, the parameters fitted by MLDS are the perceptual scale values that correspond to each stimulus level, and that collectively define the perceptual scale. MLDS essentially finds the best weights for the scale values that correspond to each stimulus level (except the first and last, which are not free parameters and are set to 0 and 1). MLDS also fits a value for the decision noise. The decision noise is the error associated with each trial decision.

As with most fitting procedures, one has to make initial guesses for the free parameters. Probably the best guess is that the perceptual scale parameters are linearly-spaced, although in many instances a compressive function such as the power function we described above will be a better guess. To make a linear scale of guesses between 0 and 1 execute the following:

```
>>PsiValuesGuess = [0:1/(NumLevels-1):1];
```

with **NumLevels** set to 10.

We are now ready to run the MLDS fitting routine **PAL_MLDS_Fit**. It has the form:

```
>>[PsiValues SDnoise LL exitflag output] = PAL_MLDS_...
Fit(StimList, NumPos, OutOfNum, PsiValuesGuess,
SDnoiseGuess);
```

The last, new, argument is the initial guess for the decision noise standard deviation (SD). You can set this again to 0.3. The function returns a vector **PsiValues** which contains the list of fitted parameters. The number of parameters in **PsiValues** corresponds to the number of stimulus levels, but remember that the first and last of these have already been set to 0 and 1. **SDnoise** is the estimate of the decision noise SD. **LL** is the log likelihood (see Section B), **exitflag** is 1 if the routine converged, 0 if it did not, and **output** is a structure that contains some details regarding the iterative search.

Finally, to obtain estimates of the errors associated with each of the estimated scale parameters, we perform a bootstrap analysis using **PAL_MLDS_Bootstrap** by typing and executing:

```
>>[SE_PsiValues SE_SDnoise] = PAL_MLDS_Bootstrap(StimList,Out...
OfNum,PsiValues,SDnoise,400);
```

PsiValues and **SDnoise** contain the values that resulted from the MLDS fit. The last parameter sets the number of bootstrap iterations.

Both **PAL_MLDS_Fit** and **PAL_MLDS_Bootstrap** use an iterative search procedure and you can deviate from the default search parameters using an "options" argument at the end of the input parameter list. Details of how to do this can be found at the end of Section 4.3.3.1.2 in Chapter 4. In **PAL_MLDS_Bootstrap** the routine might fail to fit a simulated dataset. If this is the case, the routine will issue a warning and the standard errors it returns should not be used. The problem may be helped by having the routine try the fit a few more times, starting with different initial guesses for the parameters. You can add an optional argument **'maxTries'** to the function call and the routine will try fitting any failed fits a few times using different initial values for the parameters each time. The optional argument **'maxTries'** is used in an entirely analogous fashion as it is in the function **PAL_PFML_BootstrapParametric** (Section 4.3.3.1.3). Note that fits to some simulated datasets may never converge. This might happen especially when an experiment consists of relatively few trials or when the value of **SDnoise** is high.

7.2.1.4 Plotting the Results of MLDS

First plot a green line for the Logistic function used to generate the artificial data set.

```
>>StimLevelsGenPlot = [1:0.1:9];
>>PsiValuesGenPlot = PAL_Logistic([5 1 0...
 0],StimLevelsGenPlot);
>>plot(StimLevelsGenPlot, PAL_Scale0to1(PsiValuesGenPlot),...
 'g-');
>>hold on
```

And now add in the MLDS-fitted perceptual scale values and the associated standard errors that were derived by bootstrapping:

```
>>plot(1:NumLevels, PsiValues, 'k-s');
>>for i = 2:length(SE_PsiValues)-1
line([i i],[PsiValues(i)-SE_PsiValues(i) PsiValues(i) +...
SE_PsiValues(i)], 'color','k');
end
```

The result should look something like Figure 7.2. It is very important to remember that the green line in the figure is the function used to *generate* the responses in the simulated paired comparison task and is *not* a fit to the data. The fits to the data and their bootstrap errors are the open black squares and error bars.

7.2.1.5 Running the MLDS Demonstration Program

The various steps above can be run together as a complete sequence using the following demonstration routine in Palamedes:

```
>>PAL_MLDS_Demo
```

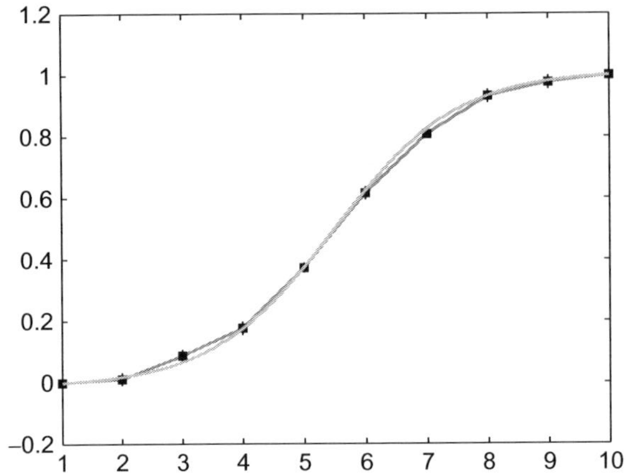

FIGURE 7.2 Example output of MLDS. The values on the abscissa are stimulus magnitudes and on the ordinate perceptual magnitudes. The green line is the function used to *generate* the hypothetical data and is *not* a fit to the data. The square symbols are the MLDS-calculated estimates of the perceptual scale values associated with each stimulus level. Error bars are standard errors derived by bootstrapping.

The script prompts the user for the type of method (pairs, triads or quadruples), the number of stimulus levels, the number of repeats of each stimulus combination, and the hypothetical observer's internal noise level. Thus, if these arguments are set to 3, 10, 30, and 0.3, the routine will output a graph that should look something like Figure 7.2. The program outputs the number of trials the experiment simulates, which for our example should be 2820. However, this number would be different if the **`MaxDiffBetweenLevels`** parameter, which is set inside the program to 3, is changed.

7.3 SECTION B: THEORY

In Section B we describe the theory behind MLDS, consider an important issue that is pivotal to evaluating the relative merits of different scaling procedures, and discuss partition scaling.

7.3.1 How MLDS Works

Let us begin with the example of the method of quadruples. Call the set of stimulus magnitudes $S_1, S_2, S_3, S_4, S_5, S_6 \ldots S_N$. Remember that on each trial four different stimulus magnitudes are presented to the observer in two pairs, and the observer decides which pair is more different (or more similar). MLDS treats the set of values

$\psi(2), \psi(3) \ldots \psi(N-1)$ as free parameters that have to be estimated; $\psi(1)$ and $\psi(N)$ are fixed at 0 and 1. Let's say on trial one, the two pairs of the quadruple are S_1S_2 and S_3S_4, and the observer responds that pair S_1S_2 is the more different. For a given test set of $\psi(S)$s, MLDS calculates the probability that a hypothetical observer characterized by these parameters will respond that S_1S_2 has the larger perceived difference. The result is the likelihood associated with that set of $\psi(S)$ for this one trial. The calculation is then repeated for the next trial, say for S_1S_6 and S_2S_4, and so on until the likelihoods of all the trials have been calculated. The likelihoods of all the trials are then multiplied to obtain the across-trials likelihood. The entire procedure is then repeated for a different test set of $\psi(S)$s. After searching through the parameter space in this way the set that gives the maximum across-trials likelihood is chosen.

Let us work through the first trial example in more detail. We start with initial guesses $\psi(1) = 0.5$, $\psi(2) = 0.7$, $\psi(3) = 0.2$, and $\psi(4) = 0.3$. Let us also assume that the internal decision noise $\sigma_d = 0.1$. Now we calculate the probability that the observer will respond "S_1S_2 more different,"*given* those values. First we compute a value D that corresponds to the difference-between-the-difference-between scale values. This is:

$$D = |\psi(2) - \psi(1)| - |\psi(4) - \psi(3)| = |(0.7 - 0.5)| - |(0.3 - 0.2) = 0.1 \qquad (7.2)$$

To convert D into a probability, it is first converted to a z-score by dividing by σ_d, which for this example results in a value of 1. The area under the normal distribution below this value is then calculated, which is 0.8413. This means that the likelihood of the response "S_1S_2 more different,"*given* the above values of $\psi(1), \psi(2), \psi(3) \ldots \psi(N)$ and *given* a noise σ_d of 0.1, is 0.8413.

Using the same set of $\psi(S)$ values and the same σ_d, the algorithm proceeds similarly to calculate the likelihoods for each of the other trials, which will include all other quadruples. On those trials in which the response to the $S_1S_2S_3S_4$ quadruple is "S_3S_4 more different" the likelihood will be $1 - 0.8413 = 0.1587$ (the two likelihoods must sum to unity). Once the likelihoods have been calculated for all trials, we multiply them out to obtain their joint probability, i.e., across-trials likelihood. However, as with the calculation of likelihoods in Chapter 4, rather than multiply out the individual likelihoods across trials and then compute the logarithm of the result, we take the logarithm of each likelihood and sum across trials. Thus:

$$LL(\psi(1), \psi(2) \ldots \psi(N), \sigma_d \,|\, \mathbf{r}) = \sum_{k=1}^{T} \log_e p(r_k \,|\, D_k; \psi(1), \psi(2) \ldots \psi(N), \sigma_d) \qquad (7.3)$$

where r_k is the response (0 or 1) and D_k the value of D on the kth trial, $\mathbf{r}$ the full set of responses across all trials, and T the number of trials. The whole procedure is then repeated for other parameter sets of $\psi(S)$ and σ_d. We then select the set that gives the largest across-trials likelihood. The result is the maximum likelihood estimates of the parameters for $\psi(1)$, $\psi(2)$ ...$\psi(N)$. These parameters then define the perceptual scale when plotted as a function of stimulus magnitude.

The same procedure applies to the methods of triads and paired comparisons, except that D for, say, the $S_1S_2S_3$ triad is given by:

$$D = |\psi(2) - \psi(1)| - |\psi(3) - \psi(2)| \quad (7.4)$$

(note that $\psi(2)$, which corresponds to the stimulus S_2, is common to both sides of the equation) and for, say, the S_1S_2 pair:

$$D = |\psi(2) - \psi(1)| \quad (7.5)$$

As elsewhere in Palamedes the search function employed to choose the set of $\psi(S)$s and σ_d that produce the greatest log likelihood is the **fminsearch** function in MATLAB®, whose operation is beyond the scope of this book.

7.3.2 Perceptual Scales and Internal Noise

Intuitively, one might think that the simplest method for constructing a perceptual scale is from JNDs, i.e., increment thresholds. The thought experiment goes something like this. Start with a low stimulus level – call this the first baseline. Measure the JND between this baseline and a higher stimulus level. Now set the second baseline to be the first baseline plus the JND, and measure a new JND. Now set the third baseline to the second baseline plus the second JND, measure the next JND, and so on. Eventually you will end up with a series of baselines separated by JNDs that span the entire stimulus range. If you were to plot a graph of equally-spaced points along the ordinate – these define the set of perceptual levels $\psi(S)$ – and their corresponding baselines on the abscissa (equivalent to integrating the JNDs) you will have a perceptual scale. This type of scale is termed a Discrimination scale (Gescheider, 1997), and the method for deriving it Discrimination or Fechnerian scaling (or Fechnerian integration). Fechner was the first to suggest that if increment thresholds obeyed Weber's Law (which states that increment thresholds are proportional to stimulus magnitude), the underlying psychophysical scale could be approximated by a logarithmic transform (Fechner, 1860/1966; Gescheider, 1997).

One obvious problem with constructing a scale of $\psi(S)$ by integrating JNDs is that the errors associated with each JND will tend to accumulate as one progresses to higher and higher stimulus levels, causing the perceptual scale values to stray increasingly from their "true" values. However, one can get round this problem by fitting a smooth function to a plot of JNDs against stimulus baselines, and deriving the perceptual scale via mathematical integration of the fitted function (e.g., Kingdom & Moulden, 1991).

There is, however, a potentially deeper problem with constructing perceptual scales from JNDs. The problem is that JNDs are not only determined by the "shape"

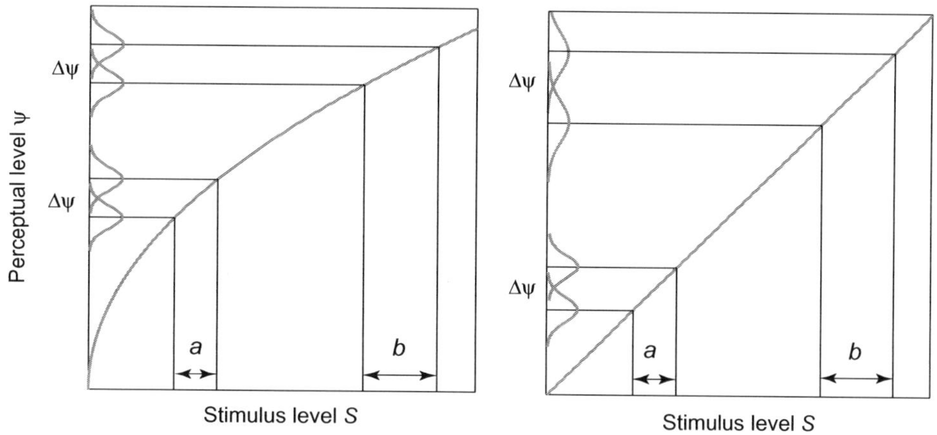

FIGURE 7.3 The impact of both the shape of the perceptual scale and the level of internal noise on JNDs. Note that the JNDs *a* and *b* are much larger than would normally be found for an abscissa that spans the full range of the stimulus dimension, for example for contrasts ranging from 0–1. See text for details.

of the perceptual scale, but also by the amount of "observer (or internal) noise" associated with each stimulus magnitude.

To understand why consider Figure 7.3. In Figure 7.3a the perceptual scale is again a power function $\psi(S) = aS^n$, where a is a (arbitrary) scaling factor and n an exponent less than 1 that determines the degree to which the function is bow-shaped; in the graph n is again 0.5. The internal noise is shown as a Gaussian distribution centered on each point on the ordinate, and as can be seen, the standard deviation, spread or dispersion of the distribution σ is the same at all points. If σ does not vary with stimulus magnitude it is often termed "additive." Formally, the addition of noise to $\psi(S)$ can be expressed by the following equation:

$$\psi(S) = aS^n + N(\sigma) \tag{7.6}$$

where $N(\sigma)$ is normally distributed noise around a mean of zero and standard deviation σ. If the noise is additive, then σ is a constant.

If one assumes that each JND, measured according to some criterion level of performance (for example 0.75 proportion correct detections), is determined by the signal-to-noise ratio $\Delta\psi/\sigma$, where $\Delta\psi$ is the corresponding average difference in internal response, then the resulting JNDs on the abscissa are a and b. Because the function is bow-shaped (or compressive), the JNDs will increase with stimulus magnitude as shown. Note that for illustrative purposes the JNDs are much larger than would be expected if one assumes that the abscissa spans the full range of the stimulus dimension.

Now consider Figure 7.3b. Here the perceptual scale is linear, not bow-shaped, and the internal noise σs are *proportional to stimulus magnitude*, which is termed "multiplicative" noise. With multiplicative noise, σ in Equation 7.6 is not a constant, but is instead proportional to stimulus magnitude, i.e., proportional to bS^n, where b scales the growth of the noise with stimulus magnitude. With multiplicative noise $\Delta\psi$ must increase with stimulus magnitude in order to maintain the criterion ratio of $\Delta\psi$ to σ. However, because the function is linear, not bow-shaped, the resulting JNDs are the same as in the figure on the left. In other words a compressive perceptual scale combined with additive noise can produce the same pattern of JNDs as a linear perceptual scale combined with multiplicative noise. It follows that it is impossible to derive the shape of the underlying perceptual scale from JNDs *unless* one knows how internal noise changes (or not) with stimulus magnitude. Put another way, if one were to assume that internal noise was additive, while in fact it was multiplicative, the perceptual scale estimated by Discrimination scaling would be invalid. Of course, internal noise is not necessarily one or the other of additive or multiplicative; it may be a combination of the two or even be non-monotonic. But the same problem applies also to these other cases.

There is an important caveat to this argument. If the purpose of a Discrimination scale is simply to define a function that predicts the pattern of JNDs, then it is by definition valid irrespective of whether the internal noise is additive or multiplicative. However, to repeat, if one wants to represent the true shape of the underlying perceptual scale, a Discrimination scale is only valid if internal noise is additive.

Is MLDS robust to whether internal noise is additive or multiplicative? Remember that the fitting procedure in MLDS not only fits $\psi(S)$ values, but also a σ for the noise associated with the decision process, i.e., an error term associated with making judgements about each pair, triad or quadruple. In reality this error term will likely be the sum of a number of different internal noise components. First, there is the internal noise associated with each point on the perceptual scale $\psi(S)$; this is the σ on the ordinate of Figure 7.3. Second, there is the internal noise associated with judging the *difference* between stimulus levels, in other words the noise associated with $\Delta\psi(S)$. With paired comparisons $\Delta\psi(S)$ is also the decision variable, whereas with triads and quadruples it is an intermediate stage of processing. Given that perceptual distance judgements tend to be Weber-like, this second noise term is likely to be proportional to $\Delta\psi(S)$, i.e., multiplicative. A third internal noise component is associated with judging the "difference-between-the-difference-between" stimulus levels, in other words the noise associated with $\Delta\Delta\psi(S)$. $\Delta\Delta\psi(S)$ is the decision variable for triads and quadruples, and again will likely be multiplicative.

The extent to which MLDS is vulnerable to incorrect assumptions about these three noise components is best answered by simulation. Our own simulations reveal that with triads and quadruples MLDS is robust to whether the internal noise levels associated with $\psi(S)$ and/or $\Delta\psi(S)$ is additive or multiplicative, provided the internal noise levels are not implausibly large. For the internal noise associated with the decision variable $\Delta\Delta\psi(S)$, Maloney and Yang (2003) have shown that with quadruples, MLDS is similarly robust to whether the noise is additive or multiplicative. Therefore, with triads and quadruples, MLDS appears to be robust to whether all three noise components are additive or multiplicative. That is, the assumption implicit in MLDS that all the noise components are additive does not result in a mis-estimation of the shape of the perceptual scale if any or all of the noise components are in fact multiplicative. On the other hand, with paired comparisons our simulations show that MLDS is not robust to whether the noise added to each $\psi(S)$ is additive or multiplicative. Therefore, we recommend that paired comparisons should only be used with MLDS if one can safely assume that the internal noise associated with each $\psi(S)$ is additive, not multiplicative.

In the next section we will argue that the partition scaling methods described in Chapter 3 are also robust to whether the noise associated with either $\psi(S)$ or $\Delta\psi(S)$ is additive or multiplicative.

7.3.3 Partition Scaling

In Chapter 3 we described various methods for partition scaling. All involved observers adjusting the magnitude of a stimulus until it was perceptually midway between two "anchor" stimuli. Here we argue that, in principle, partition scaling methods are also robust to whether the internal noise associated with each $\psi(S)$ is additive or multiplicative.

Consider Figure 7.4. In the figure, perceptual magnitude ψ, not stimulus magnitude, is shown on the abscissa. In this hypothetical example the internal noise levels are multiplicative, i.e., they increase with stimulus magnitude, as can be seen

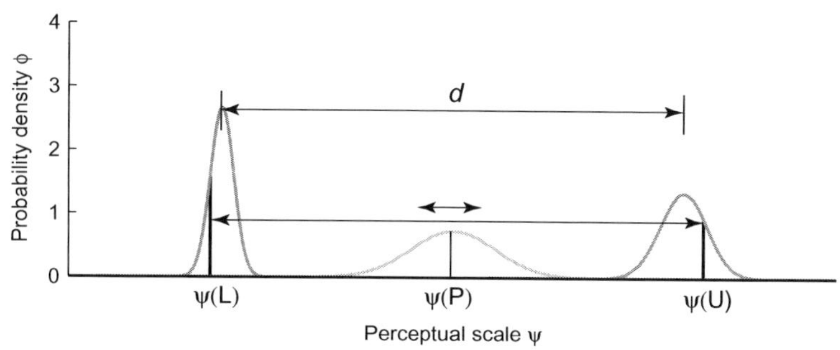

FIGURE 7.4 Effect of internal noise in a partition scaling experiment. Perceived stimulus magnitude ψ is shown on the *abscissa*. The green curves describe the distributions of ψ in response to the two anchors L (lower) and U (upper), as well as the distribution of ψ for the partition settings. d is the distance between the means of the anchor distributions.

by the different σs for the two anchor ψs. On a given trial the observer's setting will be a point midway between two random samples $\psi(L)$ and $\psi(U)$, plus an error $\psi(\varepsilon)$. This error is a combination of the internal noise associated with the partition stimulus plus a computational noise component that will likely be proportional to the perceived distance between the anchors. Thus, the distribution of $\psi(P)$ will be determined by the σs of the two anchor distributions, and a σ associated with the partition setting. Let the distance between the means of the two anchor distributions be d. If we set the mean value of the lower anchor distribution to be zero, the setting on a given trial will be:

$$\psi(P) = \frac{[\psi(L) + \psi(U)]}{2} + \psi(\varepsilon) \tag{7.7}$$

One can see intuitively from Equation 7.7 that if $\psi(L)$, $\psi(U)$, and $\psi(\varepsilon)$ are random variables from three normal distributions, the mean of $\psi(P)$ will be $d/2$, irrespective of the variance of each distribution. This follows the rule that the mean difference between two normal distributions is equal to the difference between their means, irrespective of their variances. For readers interested in a more formal proof of the idea, one needs to integrate Equation 7.7. If we denote the ordinate of a normal distribution by ϕ, then the distributions of $\psi(L)$, $\psi(U)$, and $\psi(\varepsilon)$ are given by:

$$\phi[\psi(L), \sigma_L] = \frac{1.0}{\sigma_L\sqrt{2\pi}} \exp\frac{-[\psi(L)]^2}{2\sigma_L{}^2} \tag{7.8a}$$

$$\phi[(\psi(U) - d), \sigma_U] = \frac{1.0}{\sigma_U\sqrt{2\pi}} \exp\frac{-[\psi(U) - d]^2}{2\sigma_U{}^2} \tag{7.8b}$$

and

$$\phi[\psi(\varepsilon), \sigma_\varepsilon] = \frac{1.0}{\sigma_\varepsilon\sqrt{2\pi}} \exp\frac{-[\psi(\varepsilon)]^2}{2\sigma_\varepsilon{}^2} \tag{7.8c}$$

where σ_L, σ_U, and σ_ε are the standard deviations of the lower anchor, upper anchor, and partition setting distributions. To make the following equation less cumbersome, if we denote $\psi(L)$ as l, $\psi(U)$ as u, $\psi(P)$ as p, and $\psi(\varepsilon)$ as ε, then the expected value $E(p)$ is:

$$E(p) = \int_{-\infty}^{\infty}\int_{-\infty}^{\infty}\int_{-\infty}^{\infty} \phi[l, \sigma_l] \cdot \phi[(u-d), \sigma_u] \cdot \phi[\varepsilon, \sigma_\varepsilon] \cdot \left[\frac{u+l}{2} + \varepsilon\right] dl \cdot du \cdot d\varepsilon = d/2 \tag{7.10}$$

In other words, the expected partition setting is independent of the σs associated with $\psi(L)$, $\psi(U)$, and $\psi(P)$. This will only be true, however, if the noise is symmetric, e.g., normally distributed, but this would seem to be a reasonable assumption in most cases.

In short, partition scaling should produce an interval perceptual scale that is robust to whether the internal noise levels are additive or multiplicative. By robust we do not mean unaffected. The reliability of each partition setting will be dependent on the amount of internal noise at the perceptual level of the partition stimulus, as well as the computational noise associated with making the partition judgement. However, we argue that with partition scaling the derived shape of the perceptual scale should not be *systematically* shifted from its "true" shape if internal noise is multiplicative.

To summarize: we argue that MLDS with the methods of triads or quadruples, and partition scaling methods, should result in unbiased estimates of the underlying perceptual scale. However, Discrimination scaling and MLDS using paired comparisons will only result in valid perceptual scales if one can safely assume that internal noise is additive.

Are there any advantages to partition scaling over MLDS? MLDS requires a large number of trials, especially if it uses a large number of stimulus levels, which would be necessary if the total number of discriminable steps across the stimulus range was large, as for example with contrast (e.g., Kingdom and Whittle (1996) estimated that for periodic patterns the number of discriminable steps across the full contrast range was roughly between about 40 and 90, depending on the observer and the particular stimulus). Under these circumstances partition scaling methods might prove to be more efficient.

Further Reading

An excellent and user-friendly discussion of psychological scaling procedures can be found in Chapters 9 and 10 of Gescheider (1997). A more detailed discussion can be found in the classic text on scaling by Torgerson (1958). MLDS is described in Maloney & Yang (2003). An excellent discussion on Thurstonian scaling methods can be found in McNicol (2004). Multi-dimensional scaling techniques are described in Borg & Groenen (2005).

Exercise

1. Use Palamedes to explore the relative merits of using pairs, triads and quadruples to establish a perceptual scale. Simulate experiments with pairs, triads and quadruples using the same-shaped scale for generating the hypothetical observer responses, the same number of stimulus levels, the same number of trials and the same levels of observer decision noise. Then fit the results using MLDS. Is there a difference between pairs, triads and quadruples in how close the MLDS-fitted scale values are to the generator scale? Is there a difference in the size of the bootstrap errors?

References

Borg, I., & Groenen, P. J. F. (2005). *Modern multi-dimensional scaling*. New York, NY: Springer.

Fechner, G. T. (1860/1966). Elements of psychophysics. New York, NY: Holt, Rinehart & Winston, Inc.

Gescheider, G. A. (1997). Psychophysics: The fundamentals. Mahwah, NJ: Lawrence Erlbaum Associates.

Kingdom, F., & Moulden, B. (1991). A model for contrast discrimination with incremental and decremental test patches. *Vision Research*, *31*, 851–858.

Kingdom, F. A. A., & Whittle, P. (1996). Contrast discrimination at high contrasts reveals the influence of local light adaptation on contrast processing. *Vision Research*, *36*, 817–829.

Maloney, L. T., & Yang, J. N. (2003). Maximum likelihood difference scaling. *Journal of Vision*, *3*, 573–585.

Torgerson, W. S. (1958). Theory and methods of scaling. New York, NY: Wiley.

Thurstone's, L. L. (1927). A law of comparative judgment. *Psychological Review*, *34*, 273–286.

CHAPTER

8

Model Comparisons

8.1 INTRODUCTION

As in any field of behavioral science, statistical tests are often required to make inferences about data. Ideally, psychophysical data would "speak for itself," but in reality differences between psychophysical measurements obtained under different conditions are often subtle, and the consensus (particularly strong among reviewers of research articles) is that one needs criteria to judge whether the differences are "real" or not.

The theory of statistical testing and its application to psychophysical data is an extensive and complex topic. This chapter is not intended either as a general introduction to

TABLE 8.1 Number of correct responses out of 100 trials (or 200 when combined) in a hypothetical experiment investigating whether adaptation to a stimulus affects the sensitivity to another stimulus

	Log contrast				
	−2	−1	0	1	2
No adaptation	61	70	81	92	97
Adaptation	59	59	67	86	91
Combined	120	129	148	178	188

it, or as a summary of the gamut of statistical tests available for analyzing psychophysical data. That would require a book (actually several books) in itself. Rather, we explain the logic behind the likelihood ratio test which is a statistical test which has a very general application, but we use examples taken from a particular context, namely that of testing models regarding psychometric functions. We also discuss the Palamedes toolbox routines that implement the tests. Finally, we present some alternative approaches to model selection. Much of what follows is concerned with psychometric functions, so the reader is encouraged to read at least Section A of Chapter 4 before tackling the present chapter.

Let's say you are interested in determining whether some variable X affects the performance of some task. An example would be whether adaptation, such as from prolonged viewing of a stimulus, affects the visual system's sensitivity to another stimulus. The presence or absence of adaptation would be manipulated by the researcher and is considered to be the independent variable, and you would be interested in its effects on performance in a detection task, which is considered the dependent variable. We use adaptation/no-adaption as an example of an independent variable, but many others could be used, e.g., the stimuli could be fast-moving versus slow-moving, red-colored versus green-colored, large versus small, etc. In order to determine whether an effect of adaptation exists, you measure an observer's performance twice using a 2AFC paradigm, once without adaptation and once with adaptation. You use the method of constant stimuli with 5 stimulus contrasts and 100 trials at each stimulus contrast in both conditions. The results shown in Table 8.1 are obtained. These are hypothetical data and we have chosen to use whole numbers for the log contrast values. These numbers would be unrealistic in a real experiment, but will be convenient in this and other examples.

Using the procedures discussed in Chapter 4 and implemented in the Palamedes function **`PAL_PFML_Fit`**, the conditions can be fitted individually with Logistic functions using a maximum likelihood criterion. The guessing rate parameter is fixed at 0.5, the lapse rate parameter is fixed at 0, but the threshold and slope parameters are free to vary. Figure 8.1 displays the two fitted functions as well as

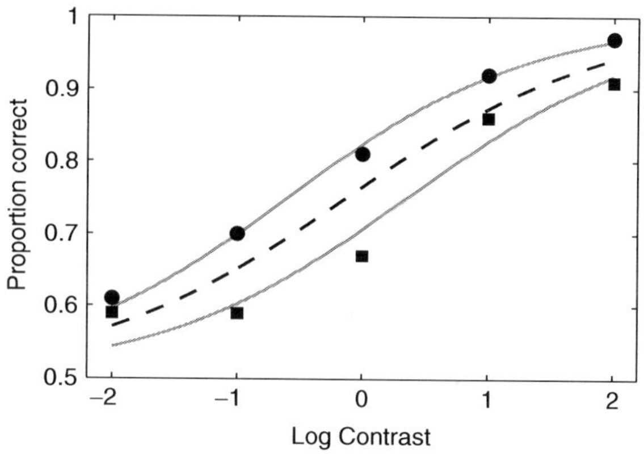

FIGURE 8.1 Proportions correct for the adaptation condition (square symbols) and no adaptation condition (round symbols), along with best-fitting Logistic functions.

the "raw" proportions correct for both of the conditions. It appears that the threshold estimates are quite different between the conditions. This is obvious from Figure 8.1 in that the fitted function for the "no adaptation" condition lies some way to the left of that for the adaptation condition. Specifically, the value for the threshold estimate in condition 1 is −0.5946, and in condition 2 it is 0.3563. The estimates for the slope parameters, on the other hand, are very close: 1.0158 and 0.9947 for conditions 1 and 2, respectively. Indeed, in Figure 8.1 the two functions appear approximately equally steep.

It is tempting to conclude from these results that adaptation indeed affects performance. After all, the two fitted functions are not identical, especially with respect to the value of the threshold parameter. The problem with that logic, however, is that this may simply be due to "sampling error." In Chapter 4 we discussed how the parameter estimates derived from experimental data are exactly that: estimates. They will not be equal in value to the "true" parameter values, but rather will vary between repeated experiments due to sampling error, even if the experiments are conducted identically. In other words, the finding that the parameter estimates in the above experiment are not identical across conditions is not a surprise at all, and does not, in and of itself, mean that the underlying, true parameters have different values. This chapter will discuss procedures that are used to answer questions about the true underlying parameter values when all we have are their estimates obtained from the limited set of data from an experiment. Such procedures are commonly referred to as inferential statistics, since they deal with making inferences about parameter values from experimental data. Note that many research questions do not concern the exact value of a threshold or slope parameter *per se*, but rather ask whether parameter values differ as a function of some independent variable. In our example above, we are not interested in the absolute level of performance with or without adaptation *per se*; rather, we are interested in whether this performance differs between the adaptation conditions.

8.2 SECTION A: STATISTICAL INFERENCE

8.2.1 Standard Error Eyeballing

In Chapter 4 we discussed the standard error of a parameter estimate. The proper interpretation of a standard error is somewhat complicated, and is discussed in Chapter 4. For our current purposes we may loosely think of a standard error as the expected difference between the true parameter value and our estimate, based on the results of our experiment. The standard errors for the parameter estimates in the above experiment were estimated using the Palamedes function **`PAL_PFML_BootstrapParametric`** (Chapter 4). Table 8.2 lists the four parameter estimates (a threshold and slope estimate for each of the two conditions) with their standard errors.

Figure 8.2 shows the threshold parameter estimates (left panel) and the slope parameter estimates (right panel) for the two conditions in the above experiment.

TABLE 8.2 Parameter estimates along with their standard errors (SE) based on the raw data shown in Table 8.1 and Figure 8.1

	Threshold	SE	Slope	SE
No adaptation	−0.5946	0.2174	1.0158	0.1814
Adaptation	0.3563	0.2207	0.9947	0.2167

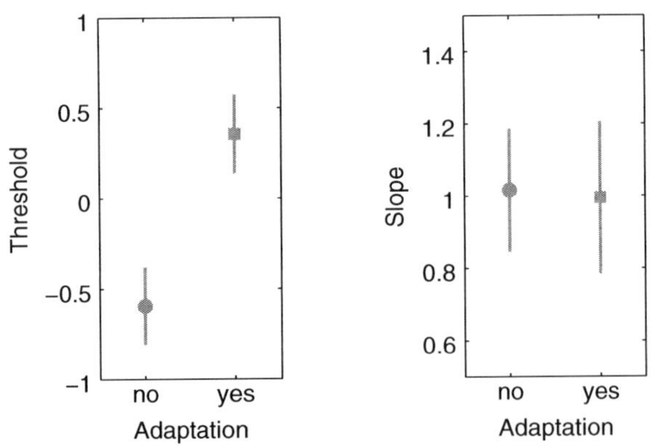

FIGURE 8.2 Graphical representation of the threshold and slope estimates and standard errors shown in Table 8.2. Standard error bars represent parameter estimate +/−1 standard error.

The vertical lines shown with each of the estimated parameters are standard error bars. Standard error bars extend from one standard error below the parameter estimate to one standard error above the parameter estimate. For example, the standard error bar of the threshold in condition 1 covers the interval -0.8120 $(-0.5946 - 0.2174)$ to -0.3772 $(-0.5946 + 0.2174)$.

The standard errors tell us something about the reliability of the parameter estimate. As a rule of thumb, we can be fairly confident that the value of the underlying true parameter will be within the range delineated by the standard error bars. Assuming the sampling distribution of parameter estimates is approximately normal in shape (a reasonable assumption in most practical situations) the standard error bars delineate the 68% confidence interval of the parameter estimate. Confidence intervals are discussed in some detail in Chapter 4. Briefly, the idea behind a confidence interval is that it makes the notion that the parameter estimate is indeed only an estimate explicit. It also expresses the degree of uncertainty regarding the value of the parameter in a manner which has some intuitive appeal. We say that we can be 68% confident that the true threshold in the no adaptation condition has a value between one standard error below its estimate and one standard error above it. Note that this does not mean that the probability that the value of the underlying parameter has a value within this range is 68%. The distinction between "probability" and "confidence" is explained in Chapter 4.

Given a graph which displays parameter estimates with their standard error bars, we can eyeball whether it is reasonable to attribute an observed difference between parameter estimates to sampling error alone. Remember that to say that the difference between parameter estimates is due to sampling error alone is to say that the parameter estimates were derived from identical true underlying parameters. Consider the left panel in Figure 8.2, which shows the threshold estimates in the two experimental conditions with their standard error bars. Given the "confidence" interpretation of standard error bars outlined above, we can be 68% confident that the true threshold in the no adaptation condition has a value within the range delineated by its standard error bar. Similarly, we can be 68% confident that the true threshold in the adaptation condition has a value within the range delineated by its standard bar. Combining these two pieces of information, it seems reasonable also to be confident that the underlying parameter values in the conditions are not equal to each other, and thus that adaptation affects performance.

A very general rule-of-thumb to adopt is to check whether the standard error bars show any overlap between conditions. If there is no overlap, as is the case for the threshold estimates in the left panel of Figure 8.2, it is usually considered reasonable to conclude that the underlying true parameters are not identical. In this example it would lead us to conclude that adaptation does increase the detection threshold. Consider now the right hand panel of Figure 8.2, which shows the estimates of the slope parameters with their standard errors for the two conditions. The slope estimates differ somewhat between the conditions, but the standard error bars show a great deal of overlap. This would lead us to conclude that the observed difference in slope estimates might very well be due to sampling error, and gives us little reason to suspect that the underlying true parameters are different between conditions.

Note that whereas it is considered acceptable to conclude that parameter values are "different" (as we did here with regard to the threshold parameters) we can never conclude that parameter values are "identical." For example, we cannot conclude that the true slope parameters are identical here; it is very possible that the slope parameter in the no adaptation condition has a true value near 1.05 and that in the adaptation condition a true value near 0.95. Note how we worded our conclusions regarding the slope parameters above. We never concluded that the slope parameters were the same, instead we concluded that the observed difference in their estimates could very well have arisen by sampling error alone. We also stated that we were given little reason to suspect that the true slope parameters were different. In other words, with respect to the slope parameters, we simply do not know whether the difference between estimates is due to sampling error, or to a difference in the underlying parameter values, or a combination of the two.

Many researchers will report the parameter estimates with their standard errors either in a table (as in our Table 8.2) or in a graph (as in our Figure 8.2) without any further statistical analysis. The reader is left to his or her own devices to draw conclusions as to whether an effect of some experimental manipulation is "real" (i.e., due to differences in the underlying true parameter values) or whether it could have resulted from sampling error alone. In many cases, it is quite clear whether one can reasonably conclude that parameters differ between conditions or whether such a conclusion is unwarranted. For example, given the results in Figure 8.2, few will argue with the conclusion that the difference in threshold estimates is probably real, whereas the difference in slope estimates could easily be attributed to sampling error alone. In other situations, it can be quite an art to eyeball whether an effect is "real" or might be due to sampling error.

Under some circumstances it might be useful to display not the standard errors of a parameter estimate in a figure, but rather some multiple of the standard error. One that is often encountered is 1.96 (or simply 2) standard error bars. In other words, the error bars extend from 1.96 standard errors below the parameter estimate to 1.96 standard errors above the parameter estimate. If we assume that the sampling distribution of the parameter estimate is normal in shape, such error bars delineate the 95% confidence interval. Within the classical hypothesis testing framework, convention allows us to conclude that the true parameter value is not equal to any particular value outside of the 95% confidence interval. So, one might prefer "1.96 standard error bars" in case one wants to show that the parameter value is different from some particular fixed value, for example the value zero in case the parameter represents a difference between conditions. Figure captions should always be clear as to what the error bars exactly represent.

8.2.2 Model Comparisons

This section describes the underlying logic behind the likelihood ratio test. This is a more formal procedure for determining whether differences between the fitted psychometric functions (PF) in different conditions are substantial enough to allow us to conclude that the parameters of the underlying true PFs are different. The problem is the same as before: even if the experimental manipulation between conditions in actuality has no effect we still expect differences in the results between the two conditions due to random factors. So the mere existence of differences in the results between conditions does not necessarily mean that the true underlying PFs are different. The logic underlying the traditional (or "frequentist" or "Fisherian") solution to this problem is the same for any statistical test you come across that results in a "p-value." This p-value is the ultimate result of any frequentist statistical test. It serves as the criterion for our decision as to whether we can reasonably conclude that an experimental manipulation affects performance. Using the example experiment described above, this section will go through one of several comparisons we might make in order to explain the concept of the p-value, and will then extend the logic to some other comparisons we could perform.

8.2.2.1 *The Underlying Logic*

We need to decide whether the observed differences among the PFs in the two conditions are real or whether they may be accounted for by sampling error alone. The specific research question is again whether adaptation affects performance on our task. Another way of looking at the question is that we aim to decide between two candidate models. One model states that adaptation does not affect performance. According to this model any observed differences in the results between the experimental conditions do not reflect a difference between the true underlying PFs, but are rather due to sampling error. In other words, performance in both conditions is governed by identical underlying PFs. Let us call this model the 1 PF model. The second model states that adaptation does affect performance. Thus, differences in the results between conditions reflect differences between conditions in the performance of the underlying sensory mechanism. There are thus two different underlying true PFs, one for each condition. Let us call this model the 2 PF model.

A different way to think about the two models is that they differ in the assumptions they make. Let us list these assumptions explicitly. The 2 PF model assumes that the probability of a correct response is constant for a given stimulus level in a given condition. What this means is that the model assumes that, as the experiment progresses, the participant does not improve or get worse (due to learning or fatigue, perhaps). We have called this assumption the "assumption of stability" in Chapter 4. It also means that the model assumes independence between trials: whether the observer gets the response on a trial correct does not affect the probability that he or she will get the response on the next trial (or any other trial) correct.

We have called this the "assumption of independence" in Chapter 4. The assumptions of stability and independence allow us to treat all 100 trials in a particular condition and at a particular contrast identically (and combine them as we did in Table 8.1 and Figure 8.1). The 2 PF model also assumes that the probability of a correct response in the no adaptation condition varies as a function of stimulus intensity in the form of a psychometric function with a particular shape (we assumed a Logistic function on log-transformed stimulus intensities). The model assumes that, in the adaptation condition also, probability correct varies with log-transformed stimulus levels according to a Logistic function. The Logistic functions in the two conditions do not necessarily have equal thresholds or slopes according to the 2 PF model. Let us further have the model assume that in both conditions the lapse rate equals 0 and the guess rate equals 0.5 (see Chapter 4).

The 1 PF model makes all the assumptions that the 2 PF model makes, but makes some additional assumptions and therefore is a bit more restrictive. The additional assumptions are that the true underlying thresholds of the PFs in both conditions are identical, and that the slopes are also identical. This is a crucial characteristic of model comparisons: one of the candidate models needs to make the same assumptions as the other model and at least one additional assumption. The statistical model comparison is used to decide whether the extra assumptions that the more restrictive model makes are reasonable. Note that when we use the term "assumption" we mean a restrictive condition. One could argue that the 2 PF model makes an assumption that the 1 PF model does not, namely that the PFs are not identical between conditions. However, if we reserve the term assumption for restrictive conditions only it is the 1 PF model which makes the additional assumption. Here, we refer to the more restrictive model as the "lesser" model and the less restrictive model as the "fuller" model.

In order to perform the statistical comparison, we start by fitting the data from both conditions twice: once under the assumptions of one of the models, and once under the assumptions of the other model. Let us first consider the 1 PF model, which claims that adaptation does not affect performance. Under the assumptions of this model, true performance is equal between conditions. In order to estimate the parameters of this single underlying function we should combine the trials across the conditions and fit a single PF to the results. Table 8.1 shows the number of correct responses combined across conditions. Of course, the number correct is now out of 200 trials per stimulus level. We can use **PAL_PFML_Fit** to fit a PF to these data using a maximum likelihood criterion. We use a Logistic function and assume a value of 0.5 for the guess rate and a value of 0 for the lapse rate. The resulting best-fitting function is shown by the broken line in Figure 8.1. It has a threshold estimate equal to −0.1251 and a slope estimate equal to 0.9544.

The 2 PF model claims that adaptation does affect performance; thus, under the assumptions of this model the underlying PFs for the two conditions will be different and we should fit each condition individually. We have already fitted this

TABLE 8.3 Model fits to experimental data and to data from the first simulation. *LR* is the likelihood ratio

	$\alpha_{\text{no adaptation}}$	$\beta_{\text{no adaptation}}$	$\alpha_{\text{adaptation}}$	$\beta_{\text{adaptation}}$	Likelihood	*LR*
Experimental data:						
1 PF:	−0.1251	0.9544	−0.1251	0.9544	1.0763×10^{-215}	0.0021
2 PF:	−0.5946	1.0158	0.3563	0.9947	5.1609×10^{-213}	
Simulation 1:						
1 PF:	−0.2441	0.9224	−0.2441	0.9224	2.0468×10^{-211}	0.6326
2 PF:	−0.2082	1.0454	−0.2768	0.8224	3.2355×10^{-211}	

model above (Section 8.1) and the fitted functions are shown in Figure 8.1 by the green lines.

Now, which is the better model? Remember that we used the "likelihood" (Chapter 4) as the metric in which to define "best-fitting." It might seem that all we need to do is determine which of the two models has the higher likelihood, and conclude it is that model which is the better one. That is a nice idea, but it will not work. To appreciate why, consider the following. Under the 2 PF model we fit the conditions separately, each with its own PF. The 2 PF model will fit identical PFs in the two conditions in the (extremely unlikely) case in which the proportions correct in condition 1 are identical to those in condition 2. In this case (and this case only), the fit of the 2 PF model would be identical to that of the 1 PF model, as would the likelihoods associated with the models. In case any difference in the pattern of results exists between the two conditions, *be it due to a real effect or sampling error*, the 2 PF model has the opportunity to fit different PFs in the two conditions in order to increase the likelihood. The 1 PF model, on the other hand, does not. It is constrained to fit a single PF to the two conditions. Thus, the 2 PF model can mimic the 1 PF model if the results in the conditions are identical, but improve its fit when the conditions are different. Another way of thinking about this is that the 1 PF model is a special case of the 2 PF model; namely the case in which the 2 PFs of the 2 PF model happen to be identical. As a result, the likelihood under the 2 PF model *will always be greater than or equal to* the likelihood of the 1 PF model.

Table 8.3 shows the parameter estimates and likelihoods for both models for this example. The likelihood under the 1 PF model is 1.0763×10^{-215}, while under the 2 PF model it is 5.1609×10^{-213}. In other words, the likelihood under the single PF model is only a fraction, equal to $1.0763 \times 10^{-215}/5.1609 \times 10^{-213} = 0.0021$, of the likelihood under the two PF model. This ratio is known as the "likelihood ratio," and is a measure of the relative fit of the two models. In cases where the results in

TABLE 8.4 Results generated by a simulated observer acting according to the 1 PF model which fits the data shown in Table 8.1 best and is displayed by the broken line in Figure 8.1

	Log contrast				
	−2	**−1**	**0**	**1**	**2**
Condition 1	61	63	77	89	96
Condition 2	59	71	75	87	94

the two conditions are exactly identical, the two model fits will also be identical, and the likelihood ratio would equal 1. In cases where the results differ between conditions, the 2 PF model will result in a higher likelihood compared to the 1 PF model and the likelihood ratio will be less than 1. The smaller the likelihood ratio, the worse is the fit of the 1 PF model relative to that of the 2 PF model.

The 1 PF model would have you believe that the relatively small value of the likelihood ratio in our experimental data can be explained entirely by sampling error. That is, according to this model the underlying true PFs in the two conditions are identical and the differences in actual observed performance between conditions are due to random factors only. The question is whether that is a reasonable explanation of the low value of the likelihood ratio. In other words, is it possible for an observer whose true PFs are identical between conditions to generate data resulting in such a low value of the likelihood ratio? One way to answer this question is simply to try it out. We simulate an observer who responds according to the 1 PF model and run this hypothetical observer many times through the same experiment that our human observer participated in. For every repetition we calculate the likelihood ratio based on the simulated results, and we see whether any of them are as small as that obtained from our human observer.

Specifically, in this situation, we test a simulated observer in an experiment that uses the same stimulus intensities as the experiment that our human observer participated in. The responses are generated in accordance with the 1 PF model which describes the performance of our human observer best (i.e., under both conditions the responses are generated by the PF shown by the broken line in Figure 8.1). Table 8.4 shows the results of the first such simulated experiment.

These simulated results are plotted in Figure 8.3 together with the best-fitting PFs under the 2 PF model (green lines) and the best-fitting PF under the 1 PF model (broken line). Table 8.3 shows the parameter estimates, likelihoods, and the likelihood ratio alongside the same information for the experimental data for both models. It is important to stress that these simulated data were generated by a hypothetical observer whose responses were known to be governed by the 1 PF model. Quite clearly, the results are much more similar between the two conditions compared to

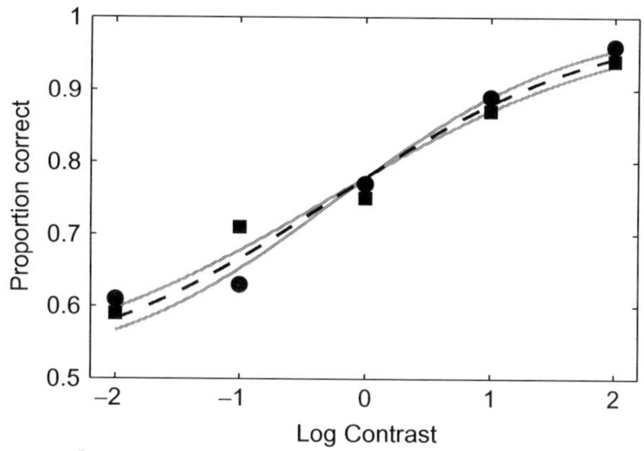

FIGURE 8.3 Data and fits of simulated experiment in which responses were generated according to the 1 PF model which fits the results of the human observer best. (i.e., the broken line in Figure 8.1).

those produced by our human observer. The separate PFs of the 2 PF model hardly differ from each other or from the single PF of the 1 PF model. Not surprisingly then, the likelihood ratio for the simulated data is 0.6326, much closer to 1 compared to the likelihood ratio we obtained from the data of our human observer. Of course, a single simulated data set resulting in a much higher likelihood ratio than our human data does not allow us to conclude much. However, we repeated the simulation a total of 10,000 times. Figure 8.4 shows the results of the first 11 of these 10,000 simulations and the best-fitting PFs to the individual conditions. The results and fits to the experimental data are also shown again. Each of the graphs also shows the corresponding likelihood ratio (*LR*).

Note from Figure 8.4 that the likelihood ratio varies systematically with the similarity between the fitted PFs in the two conditions. For example, in simulation 9 the two PFs are nearly identical and the value of the likelihood ratio is very near in value to 1. On the other hand, in simulation 3 the PFs appear quite different and the likelihood ratio is only 0.0039. However, none of the 11 simulated likelihood ratios shown in Figure 8.4 are as small as that of our experimental data. As a matter of fact, of the 10,000 simulations only 24 resulted in a smaller likelihood ratio than that based on our experimental data. Apparently it is very unlikely (24 out of 10,000 gives $p \approx 0.0024$) that an observer who acts according to the 1 PF model would produce a likelihood ratio as small or smaller than that produced by our human observer. It seems reasonable to conclude, then, that our human observer did *not* act according to the 1 PF model. The simulations indicate that we simply would not expect such a small likelihood ratio had the observer acted in accordance with the 1 PF model.

The *p*-value of 0.0024 we derived above is analogous to a *p*-value obtained from any classical, frequentist Null Hypothesis test that the reader might be more familiar with

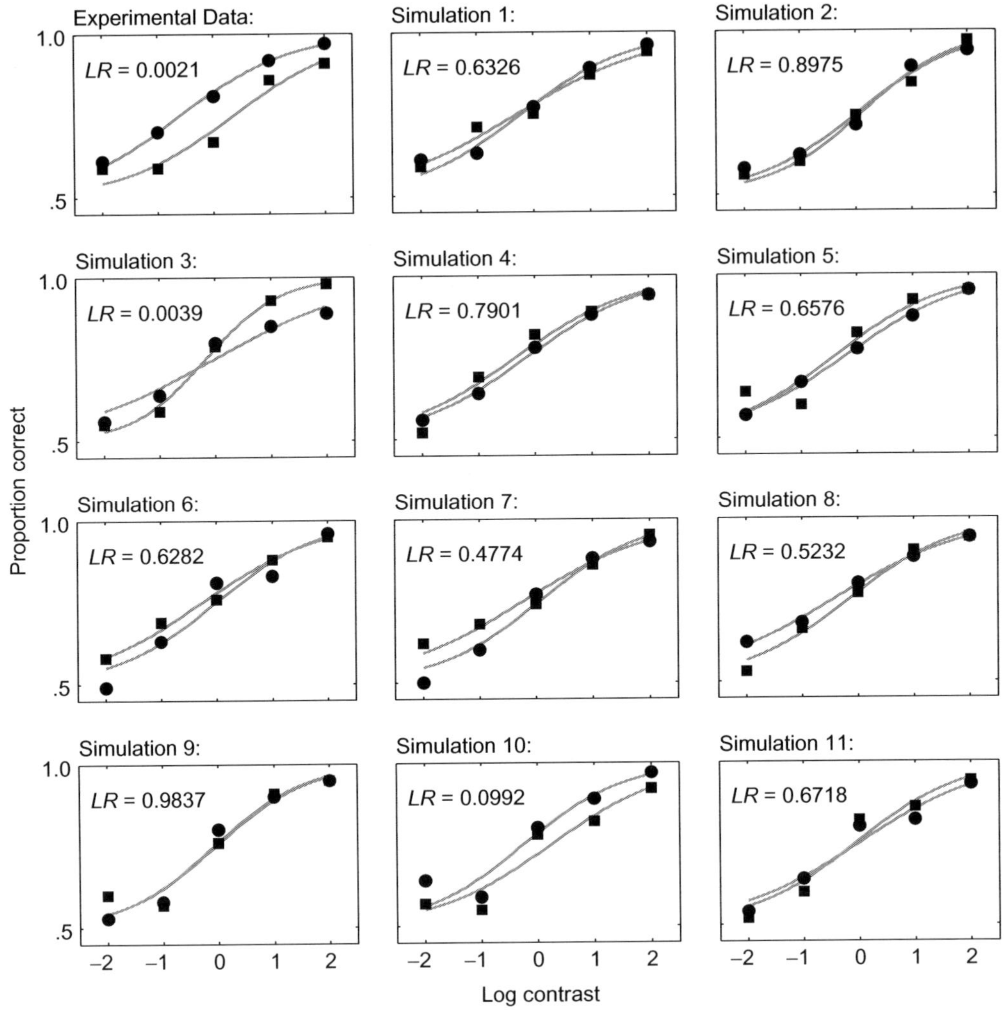

FIGURE 8.4 Experimental results as well as the results of the first 11 simulated experiments. The simulated observer acted according to the 1 PF model. Also shown are the likelihood ratios (*LR*) associated with each graph.

(be it a *z*-test, regression, ANOVA, or whatever). Any of these tests can be phrased in terms of a comparison between two models. One of the two models is always a more restrictive, "special case" version of the other. The *p*-value which results from such a test and is reported in journal articles always means the same thing. Roughly speaking, it is the probability of obtaining the observed data if the more restrictive model were true. If this probability is small (no greater than 5% by broad convention), we may conclude that the assumptions which the more restrictive model makes (but the alternative model does not) are incorrect.

We apply the same logic on a regular basis in our daily lives. Compare our logic to these two examples: "If he had remembered that today is our anniversary he would probably have said something by now. Since he has not said something by now, he has probably forgotten that today is our anniversary;" or "If it was my dog Brutus that roughed up your cat Fifi, Brutus would have probably had his face all scratched up. Since Brutus does not have his face all scratched up, it probably was not Brutus that roughed up Fifi." In our logic: "If the simpler model were true, the likelihood ratio probably would not have been as small as it was. Since it *did* come out as small as it was, the simpler model is probably not true." Note that, despite its tremendous intuitive appeal, the logic is in fact flawed, as a Bayesian thinker would be quick to point out. That is, we are making a statement about the probability that a model is true given our experimental results, but we do so based on the probability of obtaining our experimental results given that the model is true (See Chapter 4, Section 4.3.3.2.1, for a more elaborate version of the Bayesian argument).

The function **PAL_PFLR_ModelComparison** in the Palamedes toolbox is used to perform the above model comparison. We use the above example to demonstrate the use of the function. We specify the stimulus intensities in a matrix which has as many rows as there are experimental conditions, and as many columns as there are stimulus intensities in each condition. For our example we would specify:

```
>>StimLevels = [-2:1:2; -2:1:2];
```

A second matrix specifies the number of trials used at each of the stimulus levels in each condition:

```
>>OutOfNum = [100 100 100 100 100; 100 100 100 100 100];
```

A third matrix specifies the number of correct responses for each stimulus level and condition:

```
>>NumPos = [61 70 81 92 97; 59 59 67 86 91];
```

We need to specify the form of the psychometric function we wish to use as a MATLAB® inline function:

```
>>PF = @PAL_Logistic;
```

We also create a matrix that contains values for the parameter values. For both conditions we provide the initial guesses for the free parameters and specific values to use for the fixed parameters. Unless we specify otherwise, the fitting of PFs in the function will use fixed values for the guess rates and lapse rates, while the threshold and slope parameters will be free parameters.

```
>>params = [0 1 .5 0; 0 1 .5 0];
```

Finally, we create a variable which specifies the number of simulations we wish to perform to derive our statistical p-value.

```
>>B = 10000;
```

We are now ready to call our function:

```
>>[TLR pTLR paramsL paramsF TLRSim converged] =...
PAL_PFLR_ModelComparison (StimLevels, NumPos, ...
OutOfNum, params, B, PF);
```

pTLR will contain the proportion of simulated likelihood ratios that were smaller than the likelihood ratio obtained from the human data.

```
>>pTLR
pTLR =
0.0024
```

Thus, in only 24 of our 10,000 simulated experiments was the likelihood ratio smaller than that obtained from our human observer. Note that the exact value for **pTLR** might vary a bit when we run the function again, due to the stochastic nature of the simulations. **TLR** is the "transformed likelihood ratio" which is a transformation of the likelihood ratio based on the experimental data. We will have more to say about the transformed likelihood ratio in Section B of this chapter. **paramsL** and **paramsF** are the parameter estimates under the 1 PF and the 2 PF model, respectively. The L in **paramsL** stands for the lesser model, since the 1 PF model is more restrictive compared to the 2 PF model. Similarly, the F in **paramsF** stands for the fuller model.

```
>>paramsL
paramsL =
-0.1251    0.9544    0.5000    0
-0.1251    0.9544    0.5000    0
```

Note that the estimates are identical for the two conditions, which was an assumption made by the 1 PF model. Note also that these estimates are identical to those derived above by combining the results across conditions and fitting a single PF to the combined results.

```
>>paramsF
paramsF =
-0.5946    1.0158    0.5000    0
 0.3563    0.9947    0.5000    0
```

Note that these results are identical to those derived above by fitting the conditions separately (Table 8.2). Small differences might arise because of the limited precision in the search procedure that finds the maximum likelihood. In the program **PAL_PFLR_Demo** we demonstrate how to change the settings of the search algorithm in order to improve the precision. **TLRSim** is a vector of length **B** which contains the values of the transformed likelihood ratios resulting from each of the simulated experiments. **converged** is a vector of length **B** whose entries contain a 1 for each simulated experiment that was fit successfully and a 0 for each simulation that was not successfully fit. In case not all fits converged successfully, you can use the optional arguments **maxTries** and **rangeTries** in a manner analogous to their use in the functions **PAL_PFML_BootstrapParametric** and **PAL_PFML_BootstrapNonParametric** (Section 4.3.3.1.3). In this particular example, a small percentage of fits will fail on the first try and the **maxTries** and **rangeTries** options will remedy this. **PAL_PFLR_Demo** demonstrates how to use the optional arguments.

8.2.3 Other Model Comparisons

Note that the 1 PF and the 2 PF models in Section 8.2.1 differ with respect to the assumptions they make about the thresholds, as well as the slopes in the two conditions. We ended up deciding that the 2 PF model fit the data significantly better than the 1 PF model. What we do not know is whether this is because the thresholds or the slopes or perhaps both differ between conditions. The 2 PF model would also have a much better fit if, for example, only the thresholds were very different between conditions but the slopes were very similar. This, in fact, seems to be the case in the example above based on the eyeball method described in Section 8.2.1 applied to Figure 8.2. However, we may perform model comparisons that target more specific research questions.

Specifically, we can perform *any* comparison in which one of the models is a special case of the other model. In the comparison of the 1 PF and the 2 PF model we argued that the 2 PF model can always match the likelihood of the 1 PF model. It would only do so in case the results were exactly identical between the two conditions. In that case, the 2 PF would fit identical PFs to the conditions, and the likelihood under the 1 PF and 2 PF models would be identical. In case any differences exist between conditions, the 1 PF model is constrained to fit a single PF to both conditions, but the 2 PF model can accommodate the differences between conditions by fitting different PFs to the conditions. When a model is a special case of a second model, we say that it is "nested" under the second model. The likelihood ratio test is appropriate only when one of the models to be compared is nested under the alternative model.

We will now discuss two more model comparisons. Both tackle more specific research questions compared to the above 1 PF versus 2 PF comparison. Both model comparisons compare a lesser and a fuller model where the lesser model is nested under the fuller model. The first model comparison tests whether the threshold parameters differ between conditions, the second model comparison tests whether the slope parameters differ between conditions. Keep in mind that there are many more model comparisons that we could perform. As long as one of the models is nested under the other, we can use the likelihood ratio test to compare them.

8.2.3.1 *Effect on Threshold*

Perhaps you are not interested in whether adaptation has an effect on the slopes of the PF, but are only interested in the effect on the thresholds. In this case you could compare a model which assumes that thresholds and slopes differ between conditions (this is the 2 PF model from above) to a lesser model which is identical except that it constrains only the thresholds to be identical. Note that the lesser model is once again a special-case version of the fuller model. That is, under the 2 PF model it is possible to fit PFs with equal thresholds to the two conditions. Thus, the fuller 2 PF model can do everything the lesser model can do and more.

In order to perform this model comparison in Palamedes, we need to change the default settings of **PAL_PFLR_ModelComparison**. Under the default settings, **PAL_PFLR_ModelComparison** performs the test above which compares the 2 PF model to the 1 PF model. The current comparison is identical to the 2 PF versus 1 PF model test of Section 8.2.1, except that now in the lesser model only the thresholds should be constrained to be identical between conditions, but the slopes should be allowed to vary between the conditions. We can make such changes by providing **PAL_PFLR_ModelComparison** with optional arguments. These optional arguments come in pairs. The first argument in the pair indicates which setting we wish to change, and the second indicates the new value of the setting. The options are shown in Textbox 8.1.

Textbox 8.1 Defining models in **PAL_PFLR_ModelComparison**

To each of **lesserThresholds**, **lesserSlopes**, **lesserGuessRates**, **lesserLapseRates**, **fullerThresholds**, **fullerSlopes**, **fullerGuessRates**, **fullerLapseRates** the following values may be assigned:

'fixed' constrains parameter estimates in all conditions to be equal to the value specified in **params** argument.

'constrained' constrains parameter estimates in all conditions to be identical in value but this common value is a free parameter.

'unconstrained' allows each parameter estimate to take on any value.

In addition to the above three options, user may also pass a numeric array which allows for the specification of custom models (refer to Section B of this chapter).

The prefixes **'lesser'** and **'fuller'** indicate to which of the two to-be-compared models the settings should be applied.

Default settings:

```
lesserThresholds: constrained
lesserSlopes:     constrained
lesserGuesRates:  fixed
lesserLapseRates: fixed
fullerThresholds: unconstrained
fullerSlopes:     unconstrained
fullerGuesRates:  fixed
fullerLapseRates: fixed
```

Our fuller model (the 2 PF model of Section 8.2.3) corresponds to the default settings. Our lesser model, however, differs from the default lesser model in that in the default lesser model the slopes are constrained to be equal. Thus, for our lesser model we need to free the slope estimates such that they can take on different values in the two conditions. We set up our variables as before, but now we call the function as follows:

```
>>[TLR pTLR paramsL paramsF TLRSim converged] =...
PAL_PFLR_ModelComparison (StimLevels, NumPos, ...
OutOfNum, params, B, PF, 'lesserSlopes', 'unconstrained');
```

When we inspect the parameter estimates under the lesser model, we note that the threshold estimates are identical in value under this model but the slope estimates are not, as we specified:

```
>>paramsL
paramsL =
-0.1906    1.1560    0.5000    0
-0.1906    0.7824    0.5000    0
```

The value of **pTLR** is once again very small:

```
>>pTLR
pTLR = 0.0015
```

Thus, we conclude that the difference in threshold estimates between the conditions reflects a real difference in the underlying threshold parameters, and that adaptation does appear to affect the threshold.

8.2.3.2 *Effect on Slope*

Similarly, we can test whether the slopes differ significantly. Our fuller model should once again be the 2 PF model which allows thresholds and slopes to differ between conditions. Now, our lesser model should constrain the slopes, but allow the thresholds to vary between conditions. We call the function as follows:

```
>>[TLR pTLR paramsL paramsF TLRSim converged] =...
PAL_PFLR_ModelComparison (StimLevels, NumPos, ...
OutOfNum, params, B, PF, 'lesserThresholds','unconstrained');
```

When we inspect the parameter estimates of the lesser model we note that the slope estimates are indeed equal in value between the two conditions, but the thresholds are not.

```
>>paramsL
paramsL =
-0.6001    1.0071    0.5000    0
 0.3605    1.0071    0.5000    0
```

```
As expected, pTLR now has a much larger value:
>>pTLR
pTLR = 0.9337
```

Thus, a likelihood ratio as small as that of our human observer could easily have arisen by sampling error alone in case the lesser model was true; i.e., we have no reason to suspect that the slopes of the underlying true PFs differ between conditions.

So far, it appears based on the model comparisons that the assumption that the slopes are equal between conditions is reasonable. The assumption that the thresholds are equal, however, does not appear to be reasonable. Going back to our original research question, then, we may reasonably conclude that adaptation affects the threshold, but not slope, of the psychometric function.

8.2.4 Goodness-of-Fit

In the adaptation experiment example, it appears that the slopes may be equal between conditions, but that the thresholds are not. However, remember that both models in each of the comparisons made additional assumptions. The assumptions made by all of the models above is that the probability of a correct response is constant for a particular stimulus level in a particular condition (assumptions of stability and independence), and that this probability is a function of log stimulus intensity by way of the Logistic function with guess rate equal to 0.5 and lapse rate equal to 0. It is very important to note that the procedure we followed to make our conclusions regarding the equality of the threshold and slope parameters is valid only insofar as these assumptions are valid. The assumptions of stability and independence are rarely made explicit in research articles, and their validity is rarely tested. The other assumptions are often explicitly verified by way of a specific model comparison, which is commonly referred to as a "goodness-of-fit test."

Although no different from any of the tests we performed above in any fundamental sense, such a model comparison is referred to as a goodness-of-fit test for reasons we hope to make evident below. A goodness-of-fit test is performed in order to test whether a particular model provides an adequate fit to some data. We briefly mentioned goodness-of-fit tests in Chapter 4 and we are now ready to discuss them in more detail. The general logic behind a goodness-of-fit test is the same as described above. A goodness-of-fit test also compares two models. Here again, one of the models is nested under the other model.

By way of example, let us determine the goodness-of-fit of the model which, so far, appears to do a good job of fitting the data obtained in our two-condition experiment above. This is the model which assumes that the slopes are identical between conditions, but the thresholds are not. For the sake of brevity we will refer to this model as the "target model."

The target model assumes stability, independence, Logistic functions with guess rate equal to 0.5 and lapse rate equal to 0, and equal slopes between conditions. A goodness-of-fit test is used to test the validity of all these assumptions of the target model simultaneously, except for the assumptions of stability and independence. It does so by comparing the target model against a model which makes *only* the assumptions of stability and independence. The model which assumes only stability and independence is termed the "saturated model." In the saturated model, the parameters corresponding to the probabilities of a correct response are not constrained at all. That is, for each stimulus intensity in each of the conditions, the estimate of the probability of a correct response is free to take on any value entirely independent of the probabilities of correct responses at other stimulus intensities or conditions. Thus, the saturated model requires the estimation of the probability of a correct response for each particular stimulus intensity in each condition. Note that the target model is nested under the saturated model. That is, under the saturated model the probabilities of a correct response are free to take on any value, including those that would collectively conform exactly to the target model. As such, the target model could not possibly produce a better fit (as measured by the likelihood) compared to the saturated model, and thus we can perform the likelihood ratio test.

Now that we have identified our two models and made sure that one is nested under the other, we proceed exactly as we did in the tests we performed above. We simulate the experiment many times using a hypothetical observer which we programmed to respond in accordance with the more restrictive, or lesser, target model. We fit the data of each simulated experiment twice: once under the assumptions of the target model; and once under the assumptions of the saturated model. Under the saturated model, the fitting consists of finding the probability of a correct response for each condition and stimulus intensity which maximizes the likelihood. These will simply be the observed proportions of correct responses. For each simulated experiment we calculate the likelihood ratio based on these fits. If the likelihood ratio computed from our experimental data seems to be similar to those obtained from the simulated experiments, it seems reasonable to conclude that our human observer acted like the target model (i.e., the target model fits the data well). If the likelihood ratio obtained from our experimental data is much lower than those typically obtained from the simulated observer, we decide that at least one of the assumptions made by the target model, but not by the saturated model, is invalid (i.e., the target model does not fit the data well).

The Palamedes function that performs a goodness-of-fit test when we have more than one condition is **PAL_PFML_GoodnessOfFitMultiple**. We define **StimLevels**, **NumPos**, **OutOfNum**, **B**, and **PF** as above. We also need to specify the parameter values to be used while generating responses during the simulations. We use the best-fitting parameter estimates obtained from our experimental data and derived under the assumptions of the target model.

```
>>params = [-0.6001 1.0071 0.5 0; 0.3605 1.0071 0.5 0];
```

We are now ready to call the function. In the function call we specify our target model. That is, we specify whether the thresholds, slopes, guess rates, and lapse rates in the target model are constrained to be identical between conditions, are free to differ between conditions, or have a fixed value.

```
>>[TLR pTLR TLRSim converged] =...
PAL_PFML_GoodnessOfFitMultiple(StimLevels, NumPos, ...
OutOfNum, params, B, PF, 'Thresholds', 'unconstrained', ...
'Slopes', 'constrained', 'GuessRates', 'fixed', ...
'LapseRates', 'fixed');
```

After the routine completes, we can inspect the statistical *p*-value:

```
>>pTLR
pTLR =
0.9167
```

Thus, the target model provides an excellent fit to the experimental data (after all, our experimental data produced a higher likelihood ratio than 92% of the data sets that were actually produced according to the target model).

Note that when we defined the target model we did not specify it to have the particular parameter values that we estimated from our data. That is, for each of the simulations the parameter estimates of the lesser model were determined from the simulated data themselves. This is crucial, because if we were to force the lesser model for all the simulated datasets to have the specific parameter estimates that we derived from our data, our resulting *p*-value would be hard to interpret. The problem is that the model we specify should be a model of the actual underlying process. However, the parameter estimates that we derive from our data are only estimates. Their exact values are tailored to the particular set of responses we collected from our observer. If we were to test our observer again, these estimates would have different values. Thus, to include the specific values of parameter estimates in the definition of the model we wish to test is inappropriate. A general rule-of-thumb to follow is that the target model you wish to test should be specified before the experimental data are collected. Of course, before the data are collected there is no way to predict what the best-fitting estimates to the data will turn out to be. Thus, we do not specify their exact values in our target model.

The transformed likelihood ratio derived in the context of a goodness-of-fit test is known as "Deviance." It is important to keep in mind, however, that despite this difference in terminology a goodness-of-fit test is not in any fundamental sense different from any other likelihood ratio test. A goodness-of-fit test is simply a test in which the fuller model is the saturated model, which is assumption-free except for the assumptions of stability and independence.

8.2.5 More Than Two Conditions

The functions **PAL_PFLR_ModelComparison** and **PAL_PFML_GoodnessOfFitMultiple** may be used to compare models involving any number of conditions. As long as one of the models is nested under the other, the models can be compared using the likelihood ratio test. Imagine you wish to expand on the above experiment by testing how thresholds vary with the duration of adaptation. You use four different durations of adaptation period: 0, 4, 8, and 12 seconds. Thus, you now have four conditions. As before, in each condition you use stimulus contrasts −2, −1, 0, 1, and 2 (in logarithmic units). You use 150 trials at each stimulus intensity in each condition for a total of 3,000 trials (4 adaptation durations × 5 stimulus contrasts × 150 trials). Table 8.5 shows the number of correct responses for the different conditions.

One might start off by fitting a PF to each condition separately. Let us say that we are confident that fixing the PFs guess rate at 0.5 is appropriate. However, we are aware that observers on occasion will lapse. As discussed in Chapter 4 (Section 4.3.3.1.2) lapses may have a large effect on threshold and slope estimates if it is assumed that the lapse rate equals 0. Thus, we wish to make the lapse rate a free parameter. We are interested in the effect on threshold, so threshold is made a free parameter. We also wish to estimate the slope of each PF. Table 8.5 lists the parameter estimates derived by fitting the conditions individually with a Logistic function using a maximum likelihood criterion with threshold, slope, and lapse rate being free parameters. Figure 8.5 plots the observed proportions correct and the fitted PFs.

We note a few problems. One is that one of the lapse rate estimates is negative, and we know that the true lapse rate cannot be negative, so this is clearly not a very good estimate. We may of course constrain the lapse rate to be non-negative (see Chapter 4). The low slope estimate in that same condition is also a bad estimate

TABLE 8.5 Number of correct responses (of 150 trials) as a function of log contrast and adaptation duration. Also shown are parameter estimates (α: threshold, β: slope, λ: lapse rate) for individually fitted conditions

	Log contrast					**Parameter estimates**		
	−2	**−1**	**0**	**1**	**2**	α	β	λ
Adaptation duration								
0 seconds	84	92	128	137	143	−0.57	1.89	0.050
4 seconds	67	85	103	131	139	0.12	1.98	0.063
8 seconds	73	85	92	125	143	0.61	1.68	0.002
12 seconds	82	86	97	122	141	0.93	1.02	−0.089

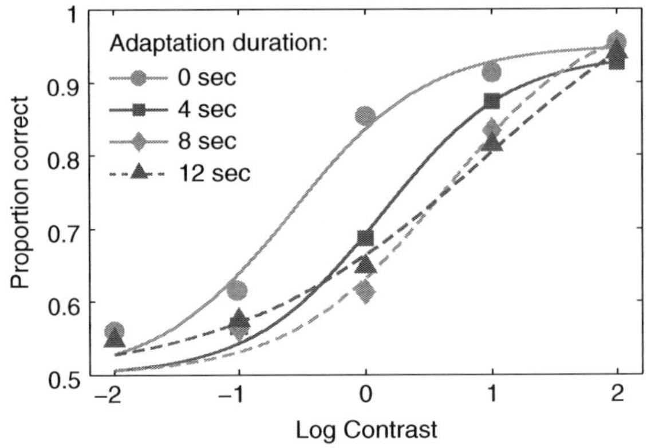

FIGURE 8.5 Proportion correct as a function of log contrast for each of four adaptation durations.

(directly related to the mis-estimate of the lapse rate). A second problem we face is that we cannot estimate standard errors by the bootstrap method because not all fits to simulated data sets will converge. The fundamental problem with our current strategy is that we are attempting to derive too many parameter estimates from too few observations. We need to decrease the number of free parameters or, alternatively, increase the number of observations (manyfold).

Let us try to reduce the number of free parameters. For example, the lapse rate may be assumed to be identical between conditions. Lapses occur when the probability of a correct response is independent of stimulus intensity (Section 4.3.1.1), for example when the stimulus presentation is missed altogether due to a sneeze and the observer is left to guess. There is little reason to suspect that lapse rates would vary with condition. Thus, we will constrain the lapse rate to be identical between conditions and effectively estimate a single lapse rate across all conditions. This has the obvious advantage that this single lapse rate will be based on four times the number of trials that a lapse rate fitted to an individual condition would be. We will also assume that the slope parameters are equal between conditions. Thus, we will constrain the slope parameter to be equal between conditions and estimate a single, shared, slope parameter. It is of course debatable whether it is reasonable to assume that slopes are equal between conditions, and we should consider whether this assumption seems reasonable on a case-by-case basis. Either way, we can test whether these kinds of assumptions are reasonable by performing a goodness-of-fit test (which we will do later).

We have now reduced the number of free parameters from 12 (4 thresholds, 4 slopes, 4 lapse rates) to 6 (4 thresholds, 1 slope, 1 lapse rate). The Palamedes function **PAL_PFML_FitMultiple** can be used to fit a model such as we have just defined to a multi-condition experiment. We need to specify a matrix containing the stimulus intensities, number of trials, and number of correct responses as we did above (Section 8.2.2.1) except, of course, that we now have four rather than two conditions.

```
>>StimLevels = [-2:1:2; -2:1:2; -2:1:2; -2:1:2];
>>OutOfNum = [150 150 150 150 150; 150 150 150 150 150; ...
150 150 150 150 150; 150 150 150 150 150];
>>NumPos = [84 92 128 137 143; 67 85 103 131 139; 73 85 ...
92 125 143; 82 86 97 122 141];
```

We also need to specify the form of PF we wish to use. As always, we define it as an inline function:

```
>>PF = @PAL_Logistic;
```

We need to specify our initial guesses for the free parameters and specify the values to be used for the fixed parameters. We define a matrix with as many rows as there are conditions in the experiment. Each row contains values for the threshold, slope, guess rate, and lapse rate, respectively.

```
>>params = [-.6 1.8 .5 .02; .1 1.8 .5 .02; .6 1.8 .5 .02; ...
.9 1.8 .5 .02];
```

Our guesses for parameter values here are guided by the fits shown in Table 8.5. We can now call our function. Within the function call, we will specify the model that we wish to apply by way of optional arguments. These arguments come in pairs (as above, Section 8.2.3) with the first argument of the pair indicating which of the parameters we wish to specify (**'Thresholds'**, **'Slopes'**, **'Guessrates'**, **'Lapserates'**) and the second indicating our assumption (**'fixed'**, **'constrained'**, **'unconstrained'**). Specifying **'fixed'** will fix the parameter value to whatever value we provided in **params**, **'constrained'** will fit a value, but it will fit the same value to all conditions, and **'unconstrained'** will fit a different value to each condition. Thus, our function call is as follows:

```
>>[paramsFitted LL exitflag output] =...
PAL_PFML_FitMultiple(StimLevels, NumPos, ...
OutOfNum, params, PF, 'Thresholds', 'unconstrained', ...
'Slopes', 'constrained', 'GuessRates', 'fixed', ...
'LapseRates', 'constrained');
```

paramsFitted will contain the estimated (or fixed) parameter values, **LL** is the log likelihood associated with the model fit, **exitflag** is set to 1 in case the fit was successful or to 0 if not, and **output** contains some information on the iterative search procedure (see description of **PAL_PFML_Fit** in Section 4.3.3.1.2). Let's inspect the parameter estimates:

```
>>paramsFitted
paramsFitted =
```

```
-0.5315  1.7412  0.5000  0.0400
 0.2144  1.7412  0.5000  0.0400
 0.4484  1.7412  0.5000  0.0400
 0.4587  1.7412  0.5000  0.0400
```

We note that, as specified, the thresholds have been allowed to take on different values between conditions, but the slopes as well as the lapse rates are identical between conditions, and the guess rates are fixed at 0.5.

Palamedes also contains functions which will perform a bootstrap analysis (Section 4.3.3.1.3) to find standard errors of the parameter estimates in a multi-condition fit. **PAL_PFML_BootstrapParametricMultiple** performs a parametric bootstrap and **PAL_PFML_BootstrapNonParametricMultiple** performs a non-parametric bootstrap. The use of both functions is demonstrated in **PAL_PFLR_FourGroup_Demo.m**, and we will demonstrate here only the use of **PAL_PFML_BootstrapParametricMultiple**. We define **StimLevels**, **OutOfNum** and **PF** as above. We also need to specify the number of simulations on which we wish to base the standard errors:

```
>>B = 400;
```

By running **PAL_PFML_FitMultiple** the best-fitting parameter estimates were assigned to **paramsFitted**, which we need to pass to the function. In the function call, we specify the model that is being considered as we did above.

```
>>[SD paramsSim LLSim converged] =...
PAL_PFML_BootstrapParametricMultiple(StimLevels, ...
OutOfNum, paramsFitted, B, PF, 'Thresholds', ...
'unconstrained', 'Slopes', 'constrained', 'GuessRates', ...
'fixed', 'LapseRates', 'constrained', 'lapseLimits',[0 1]);
```

Note that we have constrained the lapse rate to be fitted to have a positive value (Chapter 4). After the routine completes, we can inspect the standard errors:

```
>> SD
SD =
0.1476  0.2602  0  0.0166
0.1531  0.2602  0  0.0166
0.1566  0.2602  0  0.0166
0.1528  0.2602  0  0.0166
```

The **SD** matrix is organized in an identical manner to the **paramsFitted** matrix. That is, each row corresponds to a condition, and each row lists the standard error on the threshold, slope, guess rate, and lapse rate, respectively. We note that the standard errors for slopes and lapse rates are identical between the four conditions. This is because in each simulation the slopes were constrained to be identical, as were the lapse rates. The standard errors for the thresholds, on the other hand, differ between conditions. Figure 8.6 plots the threshold estimates with their standard errors.

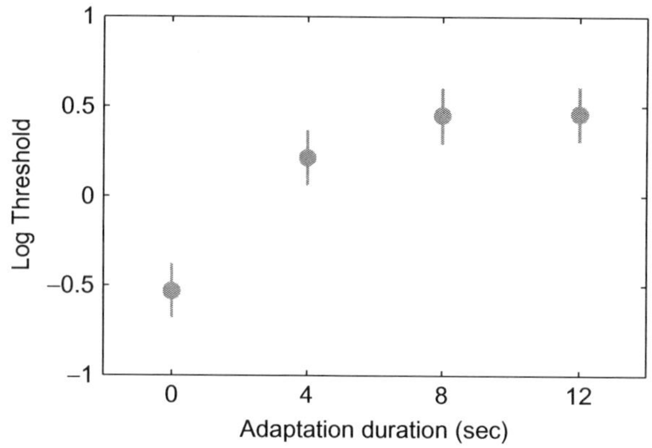

FIGURE 8.6 Plot of threshold estimates and standard errors as a function of adaptation duration (based on the hypothetical data shown in Table 8.5).

When we apply the eyeball method of Section 8.2.1 to Figure 8.6, it seems quite clear that we may reasonably conclude that the true thresholds are not equal across all four conditions. In particular, the threshold in condition 1 is very low compared to the others and, taking into consideration the standard errors, this appears to be a real effect (i.e., unlikely to occur by sampling error only).

Let us confirm this conclusion by performing a statistical model comparison. In order to do so, we need to define the appropriate lesser and fuller models. The fuller model is the model we have just fitted. Besides making the assumptions of stability and independence, it assumes that the underlying PFs are Logistic functions, that the slopes as well as the lapse rates are identical between conditions, and that the guess rate equals 0.5 for all four conditions. The lesser model is identical except that it makes the additional assumption that the thresholds are identical between conditions. Note that the lesser model is nested under the fuller model, which is a necessary condition for us to perform our model comparison by way of the likelihood ratio test.

Before we call **PAL_PFLR_ModelComparison**, we set up our arguments.

```
>>StimLevels = [-2:1:2; -2:1:2; -2:1:2; -2:1:2];
>>OutOfNum = [150 150 150 150 150; 150 150 150 150 150; ...
150 150 150 150 150; 150 150 150 150 150];
>>NumPos = [84 92 128 137 143; 67 85 103 131 139; 73 85 ...
92 125 143; 82 86 97 122 141];
>>PF = @PAL_Logistic;
>>params = paramsFitted; %assumes paramsFitted is still
%in memory. Could also have used guesses: PAL_PFLR_
%ModelComparison performs fit.
>>B = 4000;
```

We are now ready to call **PAL_PFLR_ModelComparison**:

```
>>[TLR pTLR paramsL paramsF TLRSim converged] =...
PAL_PFLR_ModelComparison (StimLevels, NumPos, ...
OutOfNum, paramsFitted, B, PF, 'lesserlapse', 'constrained', ...
'fullerlapse', 'constrained', 'fullerslope', ...
'constrained', 'lapseLimits',[0 1])
```

Remember that we only need to specify the model specifications which deviate from the default values listed in Textbox 8.1. The function will take a little while to complete (it simulates 12 million trials and performs 8,002 model fits). After it completes we inspect **pTLR**.

```
>>pTLR
pTLR = 0
```

In words, none of the 4,000 experiments that were simulated in accordance with the lesser model (in which thresholds are equal) resulted in a likelihood ratio as low as that resulting from our experimental data (i.e., $p < 1/4{,}000$). Remember that the likelihood ratio is a measure of the fit of the lesser model relative to that of the fuller model. Thus, we conclude that our experimental data were not generated by an observer acting according to the lesser model. That is, adaptation duration does appear to affect threshold.

On occasion, some of the fits to simulated datasets might not converge. Such situations might be remedied by having the routine try the fits repeatedly using initial parameter values that are randomly drawn from a range we can specify using the optional arguments **maxTries** and **rangeTries** in the same manner as we did in the function **PAL_PFML_BootstrapParametric** (Section 4.3.3.1.3). Note, however, that here also some datasets might not be fittable at all, because no local maximum exists in the likelihood function. The file **PAL_PFLR_FourGroupDemo** demonstrates the use of **maxTries** and **rangeTries**.

Note that the model comparison does not invalidate any of the assumptions that are made by both models: stability, independence, the true functions that describe probability of a correct response as a function of log contrast are Logistic functions, the slopes and the lapse rates of the true PFs are identical between conditions, and the guess rate equals 0.5. We can test all but the assumptions of stability and independence by performing a goodness-of-fit test. All arguments that need to be defined for **PAL_PFML_GoodnessOfFitMultiple** have been defined previously.

```
>>[TLR pTLR TLRSim converged] =...
PAL_PFML_GoodnessOfFitMultiple(StimLevels, NumPos, ...
OutOfNum, paramsFitted, B, PF, 'Thresholds', ...
```

```
'unconstrained', 'Slopes', 'constrained', 'GuessRates',...
'fixed', 'LapseRates', 'constrained', 'lapseLimits', [0 1]);
```

Once again, we may need the optional **maxTries** and **rangeTries** to make the fits to all simulated datasets converge. After the function completes we inspect **pTLR**:

```
>>pTLR
pTLR = 0.7218
```

In words, the data from our human observer produced a fit which was better than 72% of the data sets that were simulated in accordance with our model. Thus, we may conclude that all assumptions that our model makes, but the saturated model does not, seem reasonable. In other words, the model provides a good description of the behavior of our human observer. Once again, it is important to realize that our conclusions are only valid insofar as the assumptions that *both* models make (i.e., those of stability and independence) are valid.

Overall, then, our conclusion is that thresholds are affected by adaptation duration. Note that we may conclude only that not all underlying thresholds are equal (which is the assumption that the lesser model made but the fuller model did not). We may not conclude that they are all different. The situation is analogous to that which results when we reject the Null Hypothesis in an Analysis of Variance (ANOVA), with which the reader might be more familiar. For example, we cannot conclude that the threshold in condition 3 differs from that in condition 4. Of course, we could perform what is often termed a "pairwise comparison." This could simply be a model comparison between condition 3 and 4 disregarding the other conditions (i.e., as in the two-group comparison in Section 8.2.2). There are six such pairwise comparisons that could be performed in this four-condition experiment. A disadvantage of performing pairwise comparisons in such a manner is that we lose the distinct benefit of being able to use all data in the experiment to estimate a single lapse rate or slope. Of course, we would still be able to estimate a single slope or lapse rate across the two conditions under consideration, but we will again be pushing the number of parameters we are attempting to estimate from a relatively small amount of data.

Palamedes offers the possibility of answering more specific questions about, for example, the thresholds such as the question posed above (i.e., "do we have reason to believe that the true threshold in condition 3 might differ from that in condition 4?"), without losing the advantage of basing our lapse parameter estimate on all data collected in the experiment. Another research question that one might wish to consider is whether the decelerating nature of the thresholds as a function of adaptation duration is "real" or whether a linear trend suffices to describe the thresholds. Answering these more specific question requires a bit of technical detail and we will take up the issue and attempt to answer the above two questions in Section B of this chapter.

8.3 SECTION B: THEORY AND DETAILS

8.3.1 The Likelihood Ratio Test

All of the model comparisons in Section A were performed using what is known as the likelihood ratio test. The likelihood ratio test is a very flexible test. As long as one of the models is nested under the other and we can estimate model parameters by applying the maximum likelihood criterion, we can use the likelihood ratio test. In order to understand the details behind the likelihood ratio test, we will start off by considering a very simple example, that of coin flipping. We will then extend the logic to more complex situations.

8.3.2 Simple Example: Fairness of Coin

Let's say that we have a particular coin that we suspect is biased. Here, we consider a coin to be biased when the probability that it will land heads on any given flip does not equal 0.5. We perform a rather small-scale experiment which consists of flipping the coin ten times. The results of the ten flips are respectively (H: heads; T: tails):

HHTHTTHHTH

Thus, we obtained six heads out of ten flips. Do we have any reason to believe that the coin is not fair? In Chapter 4 (Section 4.3.3.1.1) we performed the same experiment (with the same outcome) in order to illustrate use of the likelihood function in parameter estimation. Equation 4.9 in Chapter 4 introduced the likelihood function:

$$L(a \mid \mathbf{y}) = \prod_{k=1}^{N} p(y_k \mid a) \tag{4.9}$$

where a is a potential value for our parameter of interest, $p(y_k \mid a)$ is the probability of observing outcome y on trial k *assuming* value a for our parameter, and N is our total number of trials (here, $N = 10$).

Note again that the likelihood function is a function of a. In Chapter 4 we defined the maximum likelihood estimate of α (the parameter corresponding to the probability that the coin will land heads on any flip) to be that value of a for which $L(a \mid \mathbf{y})$ attains its maximum value. For the results of the current experiment, the maximum likelihood occurs at $a = 0.6$ and this is the maximum likelihood estimate of α.

Currently, we are trying to decide whether the outcome of our experiment gives us any reason to believe that our coin is unfair. To put this a bit differently, we are trying to decide between two different models of the world. In one model the coin is fair, in the other it is not. The first model is more restrictive compared to the second because it assumes a particular value of α (0.5), whereas the second model allows α to assume any value. For this reason we refer to the models here as the lesser and fuller model, respectively. The likelihood ratio is the ratio of the likelihood under the lesser model to that of the likelihood under the fuller model, using the maximum likelihood estimate for the free parameter in the fuller model. Thus, the likelihood ratio is:

$$\Lambda = \frac{L(a = 0.5 \mid \mathbf{y})}{L(a = 0.6 \mid \mathbf{y})} = \frac{0.5^{10}}{0.6^6 \times 0.4^4} = \frac{9.766 \times 10^{-4}}{1.194 \times 10^{-3}} = 0.8176$$

The interpretation of this value is that the probability that a fair coin would produce the exact outcome of the experiment as we observed it is a fraction, equal to 0.8176, of the probability that a coin characterized by $\alpha = 0.6$ would produce the same result. Because the lesser model is a more restrictive variant of the fuller model, the likelihood ratio must have a value in the interval between 0 and 1, inclusive. In our example, it would equal 1 in case we had flipped an equal number of heads and tails, an outcome which would have given us no reason whatsoever to suspect our coin was unfair. The likelihood ratio equals 0 only when the outcome of the experiment is impossible under the lesser model, but not the fuller model. Under our lesser model, no outcome would have been impossible. Only under two possible lesser models ($\alpha = 0$ and $\alpha = 1$) would the outcome of the experiment be an impossibility.

Note that the likelihood ratio will get smaller as the proportion of flips which land heads in an experiment deviates more from the expected value under the lesser model. Specifically, Table 8.6 lists the six possible values of the likelihood ratio that may result from an experiment such as this, the outcomes that would result in these likelihoods, and the probabilities with which the six likelihood ratios would be obtained if one used a fair coin. Also listed for each possible outcome is the associated cumulative probability, i.e., the probability that a fair coin would produce a likelihood ratio equal to or smaller than that listed. For reasons to be discussed later, we often do not report the likelihood ratio, but rather a monotonic transformation of the likelihood ratio, which we refer to here as *TLR* (transformed likelihood ratio, $TLR = -2\log_e(\Lambda)$). Note that the likelihood ratio and *TLR* are functions of the results of our experiment and can thus be termed statistics. A distribution of the values of a statistic that might result from an experiment, together with the probabilities of obtaining these values and assuming a particular state of the world (here: the coin is fair or $\alpha = 0.5$) is termed a "sampling distribution" of that statistic.

As you can see from the cumulative sampling distribution of the likelihood ratio ($p(\Lambda \leq \Lambda_i \mid \alpha = 0.5)$) in Table 8.6 (and likely suspected by using common sense) obtaining a likelihood ratio as low as 0.8176 is not an unexpected outcome if the coin is fair. You would expect the likelihood ratio to be that low or lower on about

TABLE 8.6 Sampling distribution and cumulative sampling distribution of the likelihood ratio (Λ_i, i enumerates possible outcomes of experiment) and transformed likelihood ratio (*TLR*) for an experiment consisting of ten flips and assuming $\alpha = 0.5$

Number heads	Λ_i	*TLR* $[-2\log_e(\Lambda_i)]$	$p(\Lambda = \Lambda_i \mid \alpha = 0.5)$	$p(\Lambda \leq \Lambda_i \mid \alpha = 0.5)$
0 or 10	0.0010	13.8629	0.0020	0.0020
1 or 9	0.0252	7.3613	0.0195	0.0215
2 or 8	0.1455	3.8549	0.0879	0.1094
3 or 7	0.4392	1.6457	0.2344	0.3438
4 or 6	0.8176	0.4027	0.4102	0.7540
5	1	0	0.2461	1

three of every four (0.7540) similar experiments performed with a fair coin. As such, the outcome of our experiment gives us no reason to suspect that our coin is unfair. Had nine of the ten flips in our experiment come up heads, our likelihood ratio would have been 0.0252. Obtaining a likelihood as low as 0.0252 in the experiment is an unlikely thing to happen ($p = 0.0215$) when one flips a fair coin. Thus, when a coin does produce nine heads out of ten flips, it appears reasonable to doubt the fairness of the coin.

By convention, the outcome of the experiment needs to have a probability of less than 5% of occurring in case the lesser model is true before we may conclude that the lesser model is false. Note that we need to consider not the probability of the *exact* outcome. Rather, we should consider the probability of an outcome which is at least as different from that expected under the lesser model as the outcome that is actually obtained. One way in which to look at this is that we need to establish these probabilities regarding the outcome of an experiment before the experiment actually takes place. After the experiment has taken place there is no uncertainty regarding its outcome, and calculating the probability of the observed outcome is an absurdity. Before the experiment takes place we do not suspect that the coin will flip exactly nine (or any other specific number), rather we suspect it is unfair and thus expect it will flip a number of heads which is different from five.

As a bit of an aside, people nevertheless have a strong tendency to "guesstimate" the probability of seemingly improbable events that have already occurred. In case this number comes out low, they tend to rule out the possibility of the event being random and prefer other (usually incredulous) explanations. This reasoning has immense intuitive appeal, but is nevertheless invalid. Consider the following story, which is somewhat true (the details have been forgotten and made up here). One of us once witnessed a presentation by a researcher who studied twins. He related to his audience a case of two identical twins who were separated shortly after birth and grew up far apart and unaware of each other's existence. The researcher tracked down both twins and found that both twins were chiefs of their respective county's fire departments and both twins drove a 1987 blue Ford pick-up truck. The researcher mentioned that he had performed a quick casual estimate of the probability that two randomly selected individuals would both be chiefs of their respective county's fire departments and would both drive a 1987 blue Ford pick-up truck. That number was obviously extremely low. The audience was to infer, we presume, that one's choice of occupation, as well as the color, year, and make of the car one drives are genetically determined.

Had the researcher predicted beforehand that both twins would be fire department chiefs and would both drive a 1987 blue Ford pick-up truck you should be impressed. However, you should also consider that to be a rather odd prediction to make. A more sensible, but still quite odd, prediction would have been that both twins would have the same occupation (whatever it may turn out to be) and would drive similar vehicles. For the sake of argument, let's say the researcher had made this latter prediction before tracking down his separated twins. Being a much less specific prediction, the probability of it occurring by chance alone is much greater than that of a prediction which specifies particular occupations and particular vehicles. However, we imagine it would still be low enough to reject chance as an explanation if the prediction was correct (at least by the rules of classical hypothesis testing).

Unfortunately, on tracking down the twins he finds that one is an accountant who drives an older model red Geo Metro and the other is a physical therapist driving a brand new blue Nissan Pathfinder. However, as it turns out, both are passionate about building World War II model airplanes, and both own a German shepherd named Einstein. Seemingly impressive as that finding would be ("what are the odds!" right?), it is indeed only seemingly so. Apparently, we would have been impressed with the twins having any two things in common. To cover his bases then, the researcher should predict that his next pair of twins has *any* two things in common. He could then guesstimate the probability that two people have at least two things in common by pure chance alone and hope this guesstimate turns out low. He would then track down his next pair of long-separated twins and hope they have at least two things in common. If all that comes about, by the conventions of classical hypothesis testing, he could claim that chance can be ruled out as an explanation. The problem, of course, is that the probability that two random people will have at least two things in common is not low at all. We think it is quite likely, actually.

8.3.3 Composite Hypotheses

The lesser model in the above coin example states that the coin is fair, i.e., $\alpha = 0.5$. As such, the statistical properties of the coin according to the lesser model are completely specified. This allowed us to create the sampling distribution of the likelihood ratio in Table 8.6. This distribution lists all likelihood ratios that could be obtained in the experiment, and for each lists the probability with which it will be obtained in case the lesser model is true. All we needed to do to obtain this distribution was to go through all possible outcomes of the experiment, for each determine the likelihood ratio which would result from this outcome, and for each determine the probability with which it would result. The resulting sampling distribution is known as an "exact sampling distribution" (a test which uses an exact sampling distribution is known as an "exact test"). A lesser model which specifies the properties of the system completely is said to represent a "simple hypothesis."

Compare this to the example with which we started off this chapter (Section 8.2.2.1). There we wished to test whether adaptation affected sensitivity to some stimulus. The lesser model stated that adaptation does not affect sensitivity, and thus that behavior in both conditions is determined by a single underlying PF. However, it did not specify this PF completely. It did make some assumptions about the PF, namely that the shape is that of a Logistic function, that the guess rate is equal to 0.5, and that the lapse rate is equal to 0. However, it did not specify the value of the threshold parameter or the value of the slope parameter. A model which does not specify the properties of the system completely is said to represent a "composite hypothesis." In such a case, we cannot create an exact sampling distribution. In order to do so, we would have to go through all possible outcomes of the experiment (e.g., one possible outcome would be that all responses are incorrect, another would be that the response on trial 1 is correct but the responses on all other trials are incorrect, etc.). For each of these outcomes we would calculate the likelihood ratio which would result. Finally, we would have to calculate for each of these possible outcomes the probability with which the outcome would be obtained. It is the latter two that cannot be determined when the lesser model represents a composite hypothesis.

However, in case the parameter space of the lesser model is a subset of the parameter space of the fuller model, *TLR* is asymptotically distributed as the χ^2 distribution which has degrees of freedom equal to the difference in the number of free parameters in the models. A bit more formally, let $\hat{\boldsymbol{\theta}}_F$ be the maximum likelihood estimates of the parameter set $\boldsymbol{\theta}_F$ of the fuller model given observations $\mathbf{y}$. Similarly, let $\hat{\boldsymbol{\theta}}_L$ be the maximum likelihood estimates of the parameter set $\boldsymbol{\theta}_L$ of the lesser model given $\mathbf{y}$. Furthermore, let $\boldsymbol{\theta}_L \subset \boldsymbol{\theta}_F$, such that the above condition is met. The likelihood ratio is:

$$\Lambda = \frac{L(\hat{\boldsymbol{\theta}}_L \mid \mathbf{y})}{L(\hat{\boldsymbol{\theta}}_F \mid \mathbf{y})} \tag{8.1}$$

The transformed likelihood ratio (*TLR*) is given as:

$$TLR = -2 \times \log_e(\Lambda) \tag{8.2}$$

In case the lesser model is correct, *TLR* will be asymptotically distributed as χ^2 with degrees of freedom equal to the difference in number of free parameters in $\boldsymbol{\theta}_F$ and $\boldsymbol{\theta}_L$.

To be asymptotically distributed as χ^2 means that, with increasing numbers of observations, the sampling distribution of *TLR* will tend more and more towards the theoretical and continuous χ^2 distribution. Unfortunately, the number of observations that are necessary to obtain an acceptable approximation to the sampling distribution depends heavily on the particular circumstances. In many realistic settings the χ^2 approximation is quite poor (e.g., Wichmann & Hill, 2001).

An alternative is to create an empirical sampling distribution. In order to do this, we simulate the experiment many times, generating the responses in accordance with the lesser model. Of course, in order to perform the simulations we need a fully specified lesser model which includes values for the threshold and slope parameters. However, as discussed, in our example of Section 8.2.2.1, the lesser model is not fully specified. In order to be able to generate a sampling distribution, we use the maximum likelihood estimates for the free parameters (threshold and slope) that we obtained from our human observer to specify the behavior of the simulated observer completely. From each of the simulated experiments a transformed likelihood ratio (*TLR*) value is calculated in the same manner as we did for our human observer. The resulting distribution of *TLR* values will serve as our empirical sampling distribution. A distribution consisting of 10,000 simulated *TLR* values for the model comparison in Section 8.2.2.1 is shown in Figure 8.7 in the form of a histogram. Note that large *TLR* values indicate a poor fit of the lesser model compared to the fuller model. This is opposite to *LR* values where small *LR* values are indicative of a poor fit of the lesser, relative to the fuller, model. The *TLR* value obtained from our experimental data was 12.35 (indicated by the green triangle in the figure). Only 24 of the 10,000

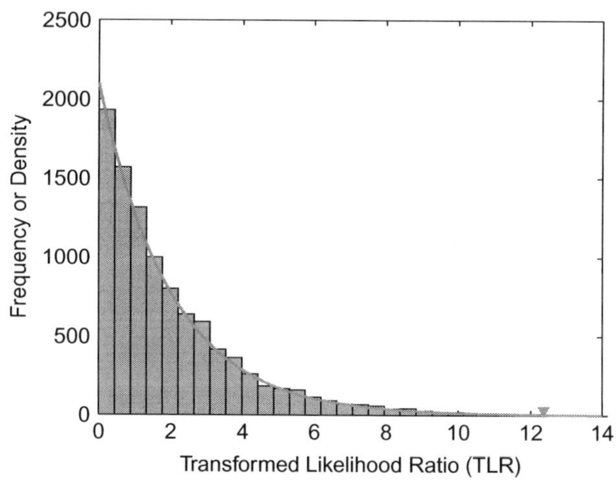

FIGURE 8.7 Empirical sampling distribution of *TLR* values for the model comparison discussed in Section 8.2.2.1, as well as the (appropriately scaled) theoretical χ^2 distribution with 2 degrees of freedom.

simulated *TLR* values were as large as that obtained from our experimental data. Thus, the human observer who generated the experimental data produced a *TLR* value which is quite unlikely to be obtained from an observer acting according to the lesser model. It seems reasonable to conclude, then, that our observer did not act according to the lesser model.

Note that the fuller model has four parameters (2 thresholds and 2 slopes) and the lesser has two parameters (1 threshold and 1 slope). Thus, the difference in the number of free parameters equals two. Also shown in Figure 8.7 is the (appropriately scaled) χ^2 distribution with 2 degrees of freedom. For this particular comparison, the $\chi^2(2)$ distribution is quite similar to our empirical sampling distribution. It may also be noted that $p(\chi^2(2) > 12.35) = 0.0021$, which is quite close to that derived from our empirical distribution (0.0024).

The value of **pTLR** which is returned by **PAL_PFLR_ModelComparison** is derived by generating an empirical sampling distribution and comparing the *TLR* of the human observer against it. The function also returns the *TLR* itself, such that the user may compare it against the χ^2 distribution with the appropriate degrees of freedom.

8.3.4 Specifying Models Using Contrasts

Linear contrasts may be used to reparameterize model parameters. This allows for a convenient and flexible manner in which to specify models. As an example, let us consider a two-condition experiment. We wish to test whether the thresholds differ between conditions. We need to define two models. In the fuller model thresholds are allowed to vary between conditions, in the lesser model they are constrained to be equal. We may use contrast matrices to transform the two thresholds into two different parameters:

$$\begin{bmatrix}\theta_1\\ \theta_2\end{bmatrix} = \begin{bmatrix}1 & 1\\ 1 & -1\end{bmatrix} * \begin{bmatrix}\alpha_1\\ \alpha_2\end{bmatrix}$$

In words, θ_1 corresponds to the sum of the thresholds, whereas θ_2 corresponds to their difference. In the fuller model, both θ_1 and θ_2 are free to vary, allowing both thresholds to take on any value. In the lesser model we make θ_1 a free parameter again, but we fix θ_2 at 0. This will constrain the two thresholds to equal each other but, as a pair, to take on any value. In other words, the model comparison can also be thought of to test whether θ_2 differs significantly from 0. In the Palamedes routines which require models to be specified (i.e., **PAL_PFML_FitMultiple**, **PAL_PFML_BootstrapParametricMultiple**, **PAL_PFML_BootstrapNonParametricMultiple**, **PAL_PFML_GoodnessOfFit** and **PAL_PFLR_ModelComparison**), models may be specified by contrast matrices. For example, in a two-condition experiment, the call

```
>>params = PAL_PFML_FitMultiple(StimLevels, NumPos, ...
OutOfNum, params, PF, 'Thresholds', 'unconstrained');
```

is equivalent to:

```
>>params = PAL_PFML_FitMultiple(StimLevels, NumPos, ...
OutOfNum, params, PF, 'Thresholds', [1 1; 1 -1]);
```

In order to fix a theta parameter to zero, we simply leave out the corresponding row of the contrast matrix. Thus, in order to constrain the thresholds to be equal between conditions we could make either of these equivalent calls:

```
>>params = PAL_PFML_FitMultiple(StimLevels, NumPos, ...
OutOfNum, params, PF, 'Thresholds', 'constrained');
>>params = PAL_PFML_FitMultiple(StimLevels, NumPos, ...
OutOfNum, params, PF, 'Thresholds', [1 1]);
```

8.3.4.1 Example: Trend Analysis

The use of contrast matrices to specify models provides for much flexibility with respect to the specific research questions that can be addressed. For example, let us revisit the four-condition experiment in Section 8.2.5, the results of which are shown in Figure 8.5. Earlier, we performed a model comparison and concluded that the assumption that the true thresholds were identical between the four conditions was untenable. This was, of course, not particularly surprising. The eyeball method leaves little doubt as to the statistical reliability of the difference between the threshold in the first condition and any of the other three thresholds. However, we may wish to answer more specific questions. For example, whereas it is quite clear that thresholds generally increase with adaptation duration, there also appears to be a decelerating trend. That is, the increase of threshold values appears to level off as adaptation duration increases. The question arises whether this effect is "real," or might be attributed entirely to sampling error.

The model comparison to be performed would involve a lesser model which does not allow thresholds to deviate from a straight line (i.e., constrains the relationship between threshold value and adaptation duration to be linear). This lesser model would be compared against a fuller model which does allow thresholds to deviate from a straight line. Readers that are well-versed in ANOVA or the general linear model will have realized by now that polynomial contrasts will allow us to formulate our models. Let us explain the logic by way of the current example. The Palamedes function **PAL_Contrasts** can be used to generate a set of polynomial contrasts. In order to do so we type:

```
>>Contrasts = PAL_Contrasts(4, 'polynomial');
```

The first argument ("**4**") specifies the number of conditions in the experiment, the second argument (**'polynomial'**) specifies that we wish to generate polynomial contrasts.

```
>>Contrasts
Contrasts =
 1.0000    1.0000    1.0000   1.0000
-1.5000   -0.5000    0.5000   1.5000
 1.0000   -1.0000   -1.0000   1.0000
-0.3000    0.9000   -0.9000   0.3000
```

Hence, the θ and α parameters are related thus:

$$\begin{bmatrix}\theta_1\\ \theta_2\\ \theta_3\\ \theta_4\end{bmatrix} = \begin{bmatrix}1 & 1 & 1 & 1\\ -1.5 & -0.5 & 0.5 & 1.5\\ 1 & -1 & -1 & 1\\ -0.3 & 0.9 & -0.9 & 0.3\end{bmatrix} * \begin{bmatrix}\alpha_1\\ \alpha_2\\ \alpha_3\\ \alpha_4\end{bmatrix}$$

Consider Figure 8.8. In panel A, all four theta parameters are fixed at 0. This will, in turn, fix all thresholds at 0. If we allow θ_1 to vary (but keep the other thetas fixed at 0), the thresholds are allowed to differ from 0, but are not allowed to differ from each other. Graphically, this would mean that all thresholds are constrained to fall on a single horizontal line (panel B). If we also allow θ_2 to vary, the thresholds are still constrained to fall on a straight line, but this line is now allowed to have a slope unequal to 0 (i.e., it is a first-order polynomial). In panel D, θ_3 is free to vary, and threshold estimates are now allowed to follow any second-order polynomial. Finally, in panel E all four θ parameters are free to vary, allowing each threshold to take on any value independent of the others.

The likelihood ratio test may be used to compare any of the models in Figure 8.8 to any other model in the figure. For any model, the parameter space is a subset of the parameter space of any model which is to its right. For example, the model comparison we performed in Section A of this chapter compared model B to model E. Based on that comparison we concluded that model B was untenable. However, the question we are currently attempting to answer requires a different comparison. The question is whether the decelerating trend in the thresholds apparent in the data is real or may have arisen due to sampling error alone. In order to answer this question we should compare model C, which assumes thresholds follow a first-order polynomial, to model D, which allows thresholds to follow a second-order polynomial which accommodates the deceleration seen in the data.

In order to perform this comparison in Palamedes we use contrast matrices to define the models. In the function **PAL_PFLR_ModelComparison** we pass the appropriate contrast matrices instead of the options **'unconstrained'**, **'fixed'**, etc. Even though it is not necessary in order to perform the model comparison, let us first fit the two models using **PAL_PFML_FitMultiple**. First we set up the necessary arrays.

Could they be, in any way, related ?

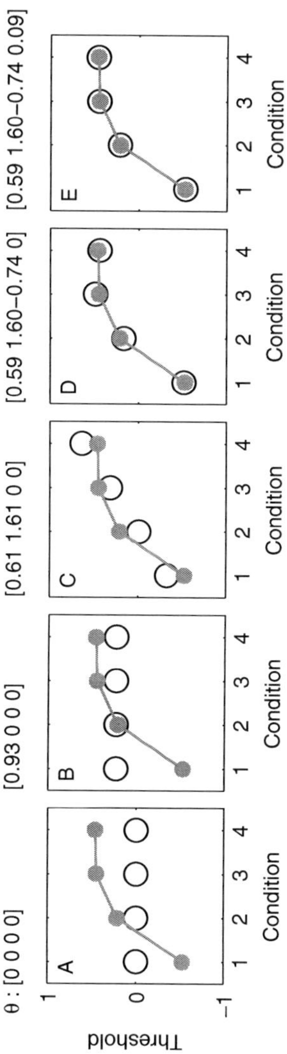

FIGURE 8.8 The results of the four-condition experiment of Section 8.2.5 (green) along with best-fitting models of varying restrictiveness (open circles).

```
>>StimLevels = [-2:1:2; -2:1:2; -2:1:2; -2:1:2];
>>OutOfNum = [150 150 150 150 150; 150 150 150 150 150; ...
150 150 150 150 150; 150 150 150 150 150];
>>NumPos = [84 92 128 137 143; 67 85 103 131 139; 73 85 ...
92 125 143; 82 86 97 122 141];
>>PF = @PAL_Logistic;
>>params = [-.6 1.8 .5 .02; .1 1.8 .5 .02; .6 1.8 .5 .02; ...
.9 1.8 .5 .02];                    %guesses
```

For each of the two models, we need to create a contrast matrix which defines it. Let us start with model C.

```
>>Contrasts = PAL_Contrasts(4, 'polynomial');
```

The full contrast matrix corresponds to model E of course, and would be equivalent to allowing thresholds to take on any value independently of the other thresholds. In order to constrain the parameters as in model C, we need to limit the contrast matrix to the first two rows, which allow the mean threshold to differ from 0, and the thresholds to increase in a linear fashion with condition, respectively.

```
>>ContrastsModelC = Contrasts(1:2,:);
```

Note that:

$$\begin{bmatrix}\theta_1\\ \theta_2\end{bmatrix} = \begin{bmatrix}1 & 1 & 1 & 1\\ -1.5 & -.5 & .5 & 1.5\end{bmatrix} * \begin{bmatrix}\alpha_1\\ \alpha_2\\ \alpha_3\\ \alpha_4\end{bmatrix} = \begin{bmatrix}\alpha_1 + \alpha_2 + \alpha_3 + \alpha_4\\ -1.5\alpha_1 - .5\alpha_2 + .5\alpha_3 + 1.5\alpha_4\end{bmatrix}$$

We are now ready to call **PAL_PFML_FitMultiple**:

```
>>paramsC = PAL_PFML_FitMultiple(StimLevels, NumPos, ...
OutOfNum, params, PF, 'Thresholds', ContrastsModelC, ...
'Slopes', 'constrained', 'GuessRates', 'fixed', 'LapseRates', ...
'constrained');
```

When we inspect **paramsC**, we note that the thresholds indeed increase linearly with condition:

```
>>paramsC
paramsC =
-0.3283   1.7138   0.5000   0.0397
-0.0071   1.7138   0.5000   0.0397
 0.3140   1.7138   0.5000   0.0397
 0.6351   1.7138   0.5000   0.0397
```

In order to define model D, we set up **ContrastsModelD** to contain the first three rows of the full contrast matrix:

```
>>ContrastsModelD = Contrasts(1:3,:);
```

And call **PAL_PFML_FitMultiple** using **ContrastsModelD** instead of **ContrastsModelC**.

```
>>paramsD = PAL_PFML_FitMultiple(StimLevels, NumPos, ...
OutOfNum, params, PF, 'Thresholds', ContrastsModelD, ...
'Slopes', 'constrained', 'GuessRates', 'fixed', ...
'LapseRates', 'constrained');
```

Let us now perform the hypothesis test. Remember that we only need to specify the deviations from the default settings which are listed in Textbox 8.1.

```
>>B = 4000;
>>[TLR pTLR paramsC paramsD TLRSim converged] =...
PAL_PFLR_ModelComparison(StimLevels, NumPos, OutOfNum, ...
paramsD, B, PF, 'lesserthreshold', ContrastsModelC, ...
'fullerthreshold', ContrastsModelD, 'lesserlapse', ...
'constrained', 'fullerlapse', 'constrained', ...
'fullerslope', 'constrained');
```

(We may again need to use the optional arguments **lapseLimits, maxTries** and **rangeTries** in order to avoid failed model fits.) On completion of the procedure, we may inspect **pTLR**:

```
>>pTLR
pTLR =
0.0065
```

Thus, the results obtained from our human observer are not likely to arise from an observer who acts strictly according to model C, and we conclude that the observed decelerating trend in thresholds is real. We leave it up to the reader to confirm that adding the fourth row in the contrast matrix (allowing all thresholds to take on any value independent of each other) does not lead to a fit which is significantly better than that of model D.

It appears that model D suffices to model the data. However, remember that the comparison between model C and model D does not in any way validate the assumptions that both models make. These are the assumptions of stability, independence of observations, the assumption that the form of the PF is a logistic function, that the slopes are equal between conditions, that the guessing rate equals 0.5, etc. We can check all but the assumptions of stability and independence by comparing model D against the saturated model, which makes only the assumptions of stability and independence.

```
>>[TLR pTLR TLRSim converged] =...
PAL_PFML_GoodnessOfFitMultiple(StimLevels, NumPos, ...
OutOfNum, paramsD, B, PF, 'Thresholds', ContrastsModelD, ...
'Slopes', 'constrained', 'GuessRates', 'fixed', ...
'LapseRates', 'constrained');
```

Upon completion, we inspect **pTLR**:

```
>>pTLR
pTLR =
0.7670
```

Thus, model D provides an excellent fit to the data.

8.3.4.2 Example: Pairwise Comparisons

Imagine that, for whatever reason, it is of interest to determine whether the difference between the thresholds in condition 3 and 4 is real or could be explained by sampling error alone. None of the comparisons performed above answers this question. Using the eyeball method with reference to Figure 8.6, few would suspect that this difference is real. However, for the sake of demonstration let us perform the comparison formally. Our fuller model should allow all thresholds (including those in conditions 3 and 4) to take on any value independent of each other, the lesser should be identical except that thresholds 3 and 4 should be constrained to equal each other. So-called Helmert contrasts allow us to define these models. The function **PAL_Contrasts** can be used to produce a set of Helmert contrasts for any number of conditions.

```
>>Contrasts = PAL_Contrasts(4, 'Helmert')
Contrasts =
1.0000     1.0000     1.0000     1.0000
0.7500    -0.2500    -0.2500    -0.2500
0          0.6667    -0.3333    -0.3333
0         -0.0000     0.5000    -0.5000
```

The first row allows the average threshold to deviate from 0. The second row allows threshold 1 to differ from the average of thresholds 2, 3, and 4. The third row allows threshold 2 to differ from the average of thresholds 3 and 4. Finally, the fourth row allows threshold 3 to differ from threshold 4. In conjunction, these contrasts allow all thresholds to take on any value, and if we use the full matrix to define the model, we will get an identical fit compared to using, for example, a full polynomial set of contrasts, or by setting the **Thresholds** option in **PAL_PFML_FitMultiple** to **unconstrained**. Let us now define the lesser model. It is the fourth row in the contrast matrix which allows thresholds 3 and 4 to differ, and this is the row we need to omit to define our lesser model.

```
>>params = repmat([0 2 .5 .02],[4 1]); %guesses
```

```
>>params = PAL_PFML_FitMultiple(StimLevels, NumPos, ...
OutOfNum, params, PF, 'Thresholds', Contrasts(1:3,:), ...
'Slopes', 'constrained', 'GuessRates', 'fixed', ...
'LapseRates', 'constrained')
params =
-0.5317    1.7412    0.5000    0.0401
 0.2142    1.7412    0.5000    0.0401
 0.4531    1.7412    0.5000    0.0401
 0.4531    1.7412    0.5000    0.0401
```

Note that the estimates of thresholds in conditions 3 and 4 are indeed equal under this model. The model comparison is performed as follows:

```
>>[TLR pTLR paramsL paramsF TLRSim converged] =...
PAL_PFLR_ModelComparison (StimLevels, NumPos, OutOfNum, ...
params, B, PF, 'lesserthreshold', Contrasts(1:3,:), ...
'fullerthreshold', Contrasts, 'lesserlapse', 'constrained', ...
'fullerlapse', 'constrained', 'fullerslope', 'constrained');
```

Inspection of **pTLR** confirms our conclusion derived by the eyeball method, which is that we have very little reason to suspect that the underlying thresholds differ between conditions 3 and 4.

```
>>pTLR
pTLR =
0.9530
```

Different research questions require different model comparisons. Contrasts may be used to define a variety of models. Trend analysis and pairwise comparisons are just two examples. For example, contrasts may also be used to test for the marginal effects of two or more variables, as well as their interaction in a factorial design (e.g., Prins, 2008). The use of contrasts to define models is routine in the context of the General Linear Model and an excellent introduction is given in Judd, McClelland, and Ryan (2008).

8.3.5 A Note on Failed Fits

On occasion, not all fits to simulated experiments converge on a solution. All Palamedes functions that perform simulations will issue a warning when this occurs. The vector **converged** returned by these functions will contain 0 for each simulation for which the fit failed and 1 for fits that were successful. We have noted above that the optional arguments **maxTries** and **rangeTries** may be used to avoid fit failures. However, this may not always make the problem go away completely. Some simulated datasets may never be fit successfully.

We encountered the problem in Chapter 4 (Section 4.3.3.1.3) and will briefly reiterate what we discussed there. The tempting solutions (ignoring the failed fits and calculating the standard error or p-value across the successful fits only, or replacing the simulations which could not be fit with new simulations, or retrying the entire set of B simulations until we have a set of B simulations which were all successfully fit) are all inappropriate, since our standard errors or p-values would be based on a non-random sample of possible simulations. Instead, we should try to make all fits successful. Generally, convergence of fits will improve with a decrease in the number of free parameters in the model(s) and with an increase in the number of observations.

If all but a very small percentage of simulations converged successfully, we might ignore the failed fits in the calculation of standard errors or p-values *as long as we report to our audience that our numbers are based on an incomplete set of simulations*. Our audience should then make up their own mind as to the value they wish to place on our results. When we are running simulations in order to estimate a p-value, we could count any unsuccessful fits as evidence contradicting the argument we wish to make. For example, assume you wish to show that adaptation affects a detection threshold. You specify a lesser model which constrains the thresholds to be equal in the adaptation and no-adaptation conditions, and a fuller model which allows thresholds to differ between the conditions. You then run **`PAL_PFLR_ModelComparison`** to derive a p-value using **`B`** = 4,000. Imagine that 103 of the simulated *TLR*s were larger than the *TLR* obtained from the experimental data, 3,882 were smaller, and the remaining 15 simulations failed to result in a succesful fit. Since you are trying to show that the lesser model is inappropriate you wish to obtain a small p-value. You could make the argument that even if these 15 failed fits would all lead to *TLR*s greater than that obtained from the experimental data your p-value would still be small enough ([103 + 15]/4,000 = 0.0295) to reject the lesser model. Once again, you would have to report that not all fits to the simulated data succeeded and you would have to report how you derived your p-value. Finally, if our purpose is to derive a p-value from a *TLR*, we might compare our *TLR* against the theoretical χ^2 distribution with the appropriate number of degrees of freedom. This does not require any simulations, of course.

8.4 SOME ALTERNATIVE MODEL COMPARISON METHODS

8.4.1 Information Criteria: AIC and BIC

The likelihood ratio test described above can only be used to compare two models in case one of the models is nested under the other. Some research questions require us to make a decision between two or more models which are not nested. In such cases, the likelihood ratio test cannot be used. Moreover, the likelihood ratio test is a "frequentist" or "Null Hypothesis" test, the fundamental logic of which is awkward at best, and for that reason rightfully disputed. The issues involved are many and we do not discuss them here (however, we have discussed what is arguably the most serious logical problem with Null Hypothesis tests in Section 4.3.3.2.1). A very readable introduction to some of the problems associated with the logic of Null Hypothesis testing is given in Cohen (1994).

This section will briefly discuss some methods that can be used to select between any two models for which a likelihood can be calculated, whether they are nested or not. Remember that using the likelihood as a metric in which to compare the goodness-of-fit of models directly is inappropriate. The reason is that this would unjustly favor models that have many parameters. For example, adding parameters to a model would *always* be preferred if we judge the fit of the model by likelihood only, because adding parameters can only increase the likelihood. However, by the scientific principle of parsimony simpler models should be preferred over more complex models. In model selection, then, the question is whether an increase in likelihood which results from the inclusion of additional parameters is worth it. Akaike's (1974) AIC (**A**n **I**nformation **C**riterion) is a measure of the relative goodness-of-fit of a model. AIC rewards increases in the likelihood, but simultaneously penalizes models for complexity (as measured by the number of free parameters included in the model). The AIC_i of any model M_i is given as:

$$AIC_i = -2LL(\hat{\boldsymbol{\theta}} \mid \mathbf{y}; M_i) + 2K_i \tag{8.3}$$

where $LL(\hat{\boldsymbol{\theta}} \mid \mathbf{y}; M_i)$ is the log likelihood for model M_i using maximum likelihood estimates for its parameters $\hat{\boldsymbol{\theta}}$, based on the observed data $\mathbf{y}$, and K_i is the number of free parameters in model M_i. The reader should not get the impression that the particular formulation of the AIC, particularly the factor 2 with which K is multiplied, is arbitrary. The derivation of equation for AIC is firmly grounded in information theory, but is well beyond the scope of this text.

Note that increases in log likelihood and decreases in the complexity of the model (both of which are to be favored) lead to smaller values of AIC_i. Thus, models with smaller associated AIC values are preferred over models with higher AIC values. Note that only the relative value of the AIC is informative as AIC is greatly dependent on the particulars of the experiment, most notably the number of trials. For this reason we should not compare AIC values between models unless the AIC values are based on the same set of observations, $\mathbf{y}$.

Let us revisit our trend analysis of Section 8.3.4.1. There we wished to determine whether the bend in the line describing thresholds as a function of adaptation duration was real or not. In order to do so, we compared model D to model C (model fits are shown in Figure 8.8). Model D allows thresholds to follow a second-order polynomial, while model C constrains them to vary according to a first-order polynomial. We performed a likelihood ratio test and concluded that the observed bend in the line would probably not have occurred in case the true trend followed a first-order polynomial. Thus, model D was preferred over model C. Let us now compare models C and D using Akaike's information criterion. The log likelihood associated with model C is -1.5542×10^3, that of model D is -1.5506×10^3 (these log likelihoods may be obtained from **`PAL_PFML_MultipleFit`**). Model C has four free parameters (a common slope, a common lapse rate, and two additional parameters to code the first-order polynomial), while Model D has five (a common slope, a common lapse rate, and three additional parameters to code the second-order polynomial).

Thus:

$$AIC_C = -2 \times -1.5542 \times 10^3 + 2 \times 4 = 3.1164 \times 10^3$$

and

$$AIC_D = -2 \times -1.5506 \times 10^3 + 2 \times 5 = 3.1112 \times 10^3$$

Model D has a lower *AIC*, and is thus preferred by this criterion. Since the *AIC* is a relative measure of fit and its absolute value is of no consequence for model selection, it is common practice to report the differences in *AIC* between models, rather than their absolute values. Table 8.7 lists, in order of fit (best to worst), the differences between the *AIC* values of all five models shown in Figure 8.8 and that of the best-fitting model (model D).

Also shown in Table 8.7 are differences between *BIC* values of the models. *BIC* stands for Bayesian Information Criterion and is given as:

$$BIC_i = -2LL(\hat{\theta} \mid \mathbf{y}; M_i) + \log_e(n)K_i \tag{8.4}$$

where the shared terms are as in *AIC*, and n is the number of observations on which the likelihood is based. In other words, the penalization for the inclusion of additional parameters increases with the sample size. We note that penalization of additional parameters is greater in *BIC* compared to *AIC* (except for the most modest

TABLE 8.7 ΔAIC and ΔBIC values for the five models shown in Figure 8.8

Model	ΔAIC	ΔBIC
D	0	0.7050
E	3.4455	10.1568
C	5.3014	0
B	26.9578	15.6500
A	29.0635	11.7494

of sample sizes: $\log_e(n) < 2$). Note that the ranking of models under the *BIC* criterion differs from that under the *AIC* criterion. The best model under the *BIC* criterion is model C, although model D (the best model under the *AIC* criterion) is a close second.

8.4.2 Bayes Factor and Posterior Odds

Model comparisons may also be performed using the Bayes Factor. The Bayes Factor gives researchers the opportunity of incorporating their prior beliefs regarding parameter values in the form of prior distributions across the parameter space. The Bayes factor is given as:

$$BF = \frac{\int L(\theta_1 \mid \mathbf{y}; M_1)\, p(\theta_1) d\theta_1}{\int L(\theta_2 \mid \mathbf{y}; M_2)\, p(\theta_2) d\theta_2} \tag{8.5}$$

where $L(\theta_i \mid \mathbf{y}; M_i)$ is the likelihood function of parameter (or parameter set) θ_i of model M_i, having observed responses $\mathbf{y}$, and $p(\theta_i)$ is the prior distribution on parameter space θ_i. The quantity $\int L(\theta_i \mid \mathbf{y}; M_i)\, p(\theta_i) d\theta_i$ is termed the marginal likelihood for model M_i. In other words, the Bayes Factor is somewhat similar to the likelihood ratio, except that it uses the mean of the likelihood function (rather than its mode as the likelihood ratio does), and it weighs the likelihood function by a prior distribution before determining its mean. A $BF > 1$ favors M_1, a $BF < 1$ favors M_2. The computation of the marginal likelihood is generally non-trivial and must, in most practical applications, occur by numerical integration.

Researchers also have the opportunity to incorporate prior beliefs regarding the relative likelihoods of the two models by applying Bayes Theorem (Section 4.3.3.2.1) to obtain the "posterior odds." Prior beliefs regarding the relative likelihoods of M_1 and M_2 are expressed as "prior odds," which are given as:

$$\textit{prior odds} = \frac{p(M_1)}{p(M_2)} \tag{8.6}$$

Note that $p(M_1)$ and $p(M_2)$ need not sum to 1. The posterior odds may be obtained by applying Bayes Theorem:

$$\textit{posterior odds} = \frac{p(M_1 \mid \mathbf{y})}{p(M_2 \mid \mathbf{y})} = BF\frac{p(M_1)}{p(M_2)} \tag{8.7}$$

Note that in case the prior probability of model M_1 equals that of M_2, the posterior odds simply equal the *BF*. Usage of Bayes Factor or posterior odds does not require that models M_1 and M_2 are nested.

Further Reading

Hays (1994) is a classic text on frequentist statistical methods. The model comparison approach emphasized in this chapter is developed much more thoroughly (but in the context of least squares error criterion methods) by Judd, McCelland, and Ryan (2008). An introduction to Bayesian statistics may be found in Jaynes (2003). Burnham and Anderson (2002) provide a thorough introduction to the information-theoretic approach to model selection which underlies *AIC* and *BIC*.

Exercises

1. A researcher conducts an experiment with two conditions. He then uses **PAL_PFLR_ModelComparison** to test whether the PFs differ between the two conditions. He assumes that a Logistic function describes the underlying mechanism well and also assumes that the lapse rate equals 0. **PAL_PFLR_ModelComparison** returns a *p*-value of 0.5604.
   a. Does this mean he can conclude that the Logistic function describes the data well?
   b. Does this mean that the lapse rate does not differ significantly from 0?
   c. What may the researcher conclude?
2. This question refers to the example data given in Section 8.2.2.1
   a. In the text (Section 8.2.3.1) it was tested whether adaptation affected the detection threshold. Repeat this test but now assume that the slopes are identical between conditions. Compare the results to that given in the text.
   b. In the text (Section 8.2.3.1) it was tested whether adaptation affected the slope parameter of the PF describing detection performance. Repeat this test but now assume that the thresholds are identical between conditions. Compare the results to that given in the text.
   c. How would you go about determining whether the assumption that the thresholds were equal between conditions is a valid assumption?
3. This question refers to the example data given in Section 8.2.5.
   a. Use contrasts to test whether the threshold at adaptation duration of 0 seconds differs significantly from that at 4 seconds.
   b. Use contrasts to test whether it is reasonable to believe that the differences that exist among the threshold estimates at adaptation durations 4, 8, and 12 seconds occurred by sampling error alone. Do this using a single model comparison only.
4. Below is a table which lists the assumptions four models make regarding the two PFs and their parameters in the two different conditions in an experiment.

	Model A	Model B	Model C	Model D	Model E
PF	Logistic	Logistic	Logistic	Gumbel	Gumbel
Thresholds	unequal	unequal	equal	equal	unequal
Slopes	unequal	equal	equal	equal	equal
Guess rate	0.5	0.5	0.5	0.5	0.5
Lapse rate	0	equal	0	equal	unequal

a. Which pairs of models may be compared using the likelihood ratio test?
b. How many free parameters does each of the models have?
c. For each of these comparisons, what may be concluded if a significant difference (i.e., $p < 0.05$) is obtained?
d. Which models may be compared against the saturated model?

5. Verify the ΔBIC values given in Table 8.7.

References

Akaik, H. (1974). A new look at the statistical model identification. *IEEE Transactions on Automatic Control, 19*, 716–723.

Burnham, K. P., & Anderson, D. R. (2002). *Model Selection and Multimodal Inference. A Practical Information-Theoretical Approach* (2nd ed.). New York, NY: Springer.

Cohen, J. (1994). The earth is round (p <0.05). *American Psychologist, 49*, 997–1003.

Hays, W. L. (1994). *Statistics*, Belmont, CA: Wadsworth Group/Thomson Learning.

Jaynes, E. T. (2003). *Probability Theory. The Logic of Science*. New York, NY: Cambridge University Press.

Judd, C. M., McClelland, G. H., & Ryan, C. S. (2008). *Data Analysis. A Model Comparison Approach*. New York, NY: Routledge.

Prins, N. (2008). Correspondence matching in long-range apparent motion precedes featural analysis. *Perception, 37*, 1022–1036.

Quick Reference Guide

Absolute threshold. Traditional term for the magnitude of a stimulus that is just discriminable from its null, as exemplified by a contrast detection threshold.
ABX. Term used in signal detection theory for a match-to-sample task with two match alternatives, i.e., the observer selects a previously viewed sample stimulus (X) from two alternative match stimuli (A,B). In this book the task is termed 2AFC match-to-sample.
Adjustment. *See* Method of adjustment.
Accuracy. Denotes how close a sensory measurement is to its corresponding physical measurement. A typical measure of accuracy is the reciprocal of the difference between the perceived and physical measurements. Example: the perceived midpoint between two dots is accurate if it is close to the physical midpoint.
Acuity. *See* Visual acuity.
Adaptive procedure. Also termed a staircase procedure. An efficient method for estimating the parameters of a PF in which the stimulus magnitude on each trial is based on the observer's responses on previous trials, such that the amount of information gained from the trial is optimized.
Additive noise. Internal noise that is constant with stimulus magnitude.
Akaike's Information Criterion (AIC). Measure of goodness-of-fit that may be used to compare the fits of two or more models to a single data set.

$$AIC_i = -2LL(\hat{\boldsymbol{\theta}} \mid \mathbf{y}; M_i) + 2K_i,$$

where $LL(\hat{\boldsymbol{\theta}} \mid \mathbf{y}; M_i)$ is the log likelihood for model M_i using maximum likelihood estimates for its parameters $\boldsymbol{\theta}$, based on the observed data $\mathbf{y}$, and K_i is the number of free parameters in model M_i. Smaller AIC values indicate better fit.
Asymmetric brightness matching. Procedure in which the observer matches the brightnesses of two stimuli set in different contexts in order to obtain their point-of-subjective-equality, or PSE.

Arcdeg. Abbreviation for arc degrees. Measure of visual angle. An arc degree is 1/360th of a full circle, or $\pi/180$ radians.
Arcmin. Abbreviation for arc minutes. Measure of visual angle. An arc minute is 1/60th of an arc degree, 1/21600th of a full circle, or $\pi/10800$ radians.
Arcsec. Abbreviation for arc seconds. Measure of visual angle. An arc second is 1/3600th of an arc degree, 1/60th of an arc minute, 1/1296000th of a full circle, or $\pi/648000$ radians.
Bayes Factor. Expresses the relative evidence for two models provided by some data.

$$BF = \frac{\int L(\theta_1 \mid \mathbf{y};M_1)p(\theta_1)d\theta_1}{\int L(\theta_2 \mid \mathbf{y};M_2)p(\theta_2)d\theta_2}$$

where $L(\theta_i \mid \mathbf{y}; M_i)$ is the likelihood function of parameter (or parameter set) θ_i of model M_i, having observed responses $\mathbf{y}$, and $p(\theta_i)$ is the prior distribution on parameter space θ_i. BF >1 favors model 1, BF <1 favors model 2.
Bayesian information criterion. Measure of goodness-of-fit that may be used to compare the fits of two or more models to a single data set.

$$BIC_i = -2LL(\hat{\boldsymbol{\theta}} \mid \mathbf{y};M_i) + \log_e(n)K_i,$$

where $LL(\hat{\boldsymbol{\theta}} \mid \mathbf{y}; M_i)$ is the log likelihood for model M_i using maximum likelihood estimates for its parameters $\boldsymbol{\theta}$, based on the observed data $\mathbf{y}$, n is the number of observations on which the likelihood is based and K_i is the number of free parameters in model M_i. Smaller BIC values indicate better fit.
Bayes' Theorem. A general statement of Bayes' Theorem is

$$p(H \mid D) = \frac{p(H)p(D \mid H)}{p(H)p(D \mid H) + p(\overline{H})p(D \mid \overline{H})} = \frac{p(H)p(D \mid H)}{p(D)}$$

where $p(H)$ is the prior probability of hypothesis H, $p(D \mid H)$ is the probability of obtaining data D assuming hypothesis H (i.e., the likelihood), and $p(H \mid D)$ is the posterior probability of hypothesis H. Bayes' Theorem allows us to adjust our prior beliefs regarding H based on our empirical results D.
Best PEST. Adaptive method for estimating a threshold. On each trial a maximum likelihood estimate is made of the threshold using the responses from previous trials and assuming a particular shape of psychometric function. The stimulus magnitude on the subsequent trial is then set to the threshold estimate.
Bias (of estimate). The difference between a parameter's true value and the expected value of its estimate.
Bias (of observer). Observer bias has two related meanings. (1) In performance tasks the tendency to make more of one type of response than another. For example in a two-interval forced-choice (2IFC) task the observer might be biased towards responding "first interval"

even if the target was equally likely to occur in both intervals. (2) In appearance tasks observer bias can refer to the difference between the point of subjective equality and the point of physical equality. For example in a vernier alignment task the point of subjective alignment might be biased away from the point of physical alignment.

Binocular rivalry. The phenomenon in which stimuli presented to the two eyes alternate in perceptual dominance.

Binomial coefficient. Formula for calculating the total number T of unique combinations of N different events, with k different events per combination.

$$T = \frac{N!}{k!(N-k)!}$$

Note that $N! = N \times (N-1) \times (N-2) \times \ldots 1$ and that 0! equals 1 by definition. For example if you have $N = 5$ stimulus magnitudes, with $k = 3$ different stimulus magnitudes presented per trial, there are a total of $T = 10$ unique combinations of stimulus magnitudes.

Bisection task. Task to measure the perceptual midpoint between two stimuli that lie at different points along a stimulus dimension. Examples: to measure the perceived midpoint of a line, or the midpoint in perceived contrast between two different contrasts.

Bisection scaling. *See* Partition scaling.

Brightness. The perceptual correlate of luminance, or light intensity.

Bootstrap method. Method used to estimate a parameter's sampling distribution through repeatedly simulating an experiment using known or assumed parameter values. The empirical sampling distribution is then used to determine the standard error of estimate of the parameter.

Cancellation procedure. *See* Nulling procedure.

Chromaticity. Specification of the color of an object regardless of its luminance, referring to both the colorfulness (or saturation) and hue.

Constant noise. *See* Additive noise.

Contrast. A measure of the relative luminance between two stimuli. Measures of contrast include Weber contrast, Michelson contrast, and RMS (root mean square) contrast. The contrast of an object with its surround is invariant to changes in the intensity of illumination.

Contrast threshold. The amount of luminance contrast required to reach a particular criterion detection performance.

Contrast sensitivity. The reciprocal of contrast threshold.

Class A observation. Term coined by Brindley (1970) for the psychophysical observation in which two physically different stimuli are perceptually indiscriminable.

Class B observation. Term coined by Brindley (1970) for the psychophysical observation in which two physically different stimuli remain perceptually discriminable even when matched along one or more stimulus dimensions.

Criterion. Usually denotes the bias of an observer towards making one type of response over another in a psychophysical task.

Criterion C. Measure of bias in a forced-choice experiment derived by signal detection analysis. For one-alternative-forced-choice (1AFC) tasks C is defined as:

$$C = -[z(pH) + z(pF)]/2$$

where $z(pH)$ and $z(pF)$ are the z-values calculated for the proportion of hits and false alarms respectively.

Criterion C′. Measure of bias in a forced-choice experiment derived by signal detection analysis:

$$C' = \frac{-[z(pH) + z(pF)]}{2[z(pH) - z(pF)]}$$

where $z(pH)$ and $z(pF)$ are z-values for the proportion of hits and false alarms respectively.

Criterion lnβ. Measure of bias in a forced-choice experiment derived by signal detection analysis, defined as:

$$\ln\beta = \ln\frac{\phi[z(pH)]}{\phi[z(pF)]}$$

where $\phi[z(pH)]$ and $\phi[z(pF)]$ are the ordinate values of a standardized normal distribution corresponding to the z-values for the proportions of hits and false alarms respectively.

Criterion-free. A psychophysical task or procedure in which observers are unlikely to be biased towards making one type of response over another, or a psychophysical measurement that is provided by an unbiased observer or computed in such a way as to take into account bias.

Criterion-dependent. A psychophysical task or procedure in which observers are likely to be biased towards making one type of response over another, or a psychophysical measurement provided by a biased observer.

Cross-modal matching. Method for measuring the apparent magnitude of a stimulus by matching it to the magnitude of a stimulus in another sensory modality. For example, the perceived slant of a visual texture might be measured by hand-adjusting the slant of an object with a planar surface.

Cumulative normal function.

$$F_N(x;\alpha,\beta) = \frac{\beta}{\sqrt{2\pi}}\int_{-\infty}^{x}\exp\left(-\frac{\beta^2(x-\alpha)^2}{2}\right)$$

The cumulative normal function is the integral of the normal distribution, where α determines the location (threshold) and β determines the slope of the function.

Detection. Usually refers to tasks/procedures/experiments that measure the magnitude of a stimulus that can be discriminated reliably from its null. Examples: measurement of a contrast detection threshold in which the null stimulus is a blank field; detection of curvature in which the null stimulus is a straight line. The term is also used to denote the measurement itself, e.g., a detection threshold.

Deviance. Term that is used for the transformed likelihood ratio (*TLR*) when the fuller model is the saturated model. Deviance is used to test the goodness-of-fit of the lesser model.

Difference threshold. Traditional term for the magnitude of a stimulus difference that is just detectable when both stimuli are above detection threshold.

Differencing strategy. Strategy of basing forced-choice decisions on the perceived differences between stimuli. Differencing strategies have been suggested to mediate perceptual decisions in the standard 2AFC, same-different, oddity, and match-to-sample tasks.

Discrimination. Most commonly refers to tasks/procedures/experiments that determine the just noticeable difference (JND) in stimulus magnitude between two stimuli with non-zero magnitude. Also used to denote the type of measurement itself, e.g., a discrimination threshold.

Discrimination scale. A perceptual scale derived by integrating JNDs. Also termed a Fechnerian scale.

Discriminand. One of the stimuli in a discrimination experiment.

d' **(d-prime).** Measurement of observer sensitivity or stimulus discriminability derived from signal detection theory. If the stimuli are assumed to be represented internally as random variables drawn from normal distributions with given means and variances, d' is a measure of the distance between the means of the distributions normalized to their standard deviation.

Equisection scaling. *See* Partition scaling.

Exposure duration (of stimulus). *See* Stimulus exposure duration.

Fechnerian integration. The method of deriving a perceptual scale by adding up or integrating discrimination thresholds (or JNDs).

Fechnerian scaling. *See* Discrimination scale.

False alarm. Responding that a stimulus is present when it is not.

Fixed parameter. A model parameter that is not allowed to vary during model fitting.

Forced-choice. Term used here to refer to any task/procedure/experiment in which the observer is required on each trial to make a response from a predefined set of choices. In the signal detection literature the term tends to be used more restrictively to refer to tasks in which the observer selects a target from two or more stimulus alternatives.

Free parameter. A model parameter that is varied during model fitting in order to optimize the fit of the model.

Geometric mean. The antilog of the mean of the logarithm of a set of numbers. If the numbers are $X_1, X_2 \ldots X_n$, the geometric mean computed using logarithms to the base 10 is given by:

$$10 \wedge \left[\frac{\sum_{i=1}^{N} \log X_i}{N} \right]$$

Geometric series. A series of numbers in which adjacent pairs have identical ratios.
Goodness-of-fit test. A statistical model comparison between two models in which the fuller model is the saturated model. The saturated model makes the assumptions of stability and independence only. As such a goodness-of-fit test tests all the assumptions of a model, except for the assumptions of stability and independence, simultaneously.
Grating induction. The illusory brightness modulation observed in a uniform stripe running at right-angles to the bars of a real luminance grating.
Guess rate. Corresponds to chance level performance: the expected proportion correct for a hypothetical observer who guesses on each trial. Note that the term is based on assumptions of the discredited high-threshold theory (under the framework of SDT an observer never truly guesses). Guess rate is the parameter of a PF that corresponds to the lower asymptote of a psychometric function (γ).
Gumbel function.

$$F_G(x; \alpha, \beta) = 1 - \exp\left(-10^{\beta(x-\alpha)}\right)$$

where α determines the location (threshold) and β determines the slope of the function. The Gumbel function is the analog of the Weibull function when a log transform on x is used and, for that reason, is sometimes referred to as the log-Weibull function or simply, but confusingly, as the Weibull function.
High-threshold Theory. A theory of detection which states that detection occurs only when the sensory evidence exceeds an internal criterion or threshold. The threshold is set such that it will not be exceeded in the absence of a stimulus (i.e., by noise alone). While these central tenets of high-threshold threshold theory have been discredited, many of the terms used in psychophysics (e.g., 'threshold', 'guess rate') are remnants of high-threshold theory.
Hit. Responding that a stimulus is present when it is present.
Hyperbolic secant function.

$$F_{HS}(x;\alpha, \beta) = \frac{2}{\pi}\tan^{-1}\exp\left(\frac{\pi}{2}\beta(x - \alpha)\right)$$

where α determines the location (threshold) and β determines the slope of the function.
Identification. Sometimes used as an alternative term to "discrimination," especially when the observer has not only to detect a stimulus, but also has to identify some additional stimulus property, such as whether the stimulus is red or green, moving leftwards or rightwards, behind or in front. Sometimes also used instead of the term "recognition."
Independence, assumption of. In the context of model fitting, this assumption states that the probability of observing a particular response ("yes," "first interval," etc.) on any given trial is not affected by observations made on other trials.
Independent observation strategy. Observer strategy of basing a forced-choice decision on the independent assessment of the likelihood that each observation is from a particular stimulus. Independent observation strategies have been suggested to underlie same-different, oddity and match-to-sample tasks.

Internal noise. The random fluctuation in the observer's internal representation of a stimulus magnitude.

Interval scale. A perceptual scale in which the differences in scale values are proportional to perceived differences in stimulus magnitude. An interval scale can be rescaled by $aX + b$ without loss of information where a and b are arbitrary constants.

Inter-stimulus-interval (ISI). The temporal interval between the offset of one stimulus and the onset of another.

Inter-trial-interval (ITI). The temporal interval between the end of one trial and the beginning of the next trial.

Just noticeable difference (JND). The smallest difference in stimulus magnitude that is just discriminable.

Lapse rate. The probability of an incorrect response that is independent of the stimulus. Lapses are most evidenced by incorrect responses to stimulus magnitudes that are considerably above threshold. Lapse rate is the parameter of a PF (λ) that determines the upper asymptote ($1 - \lambda$).

Lightness. The perceptual correlate of the reflectance, or the perceived "shade-of-gray" of an object.

Likelihood. The probability with which a hypothetical observer characterized by assumed model parameters would reproduce exactly the responses of a human observer. The likelihood is a function of parameter values, not responses. The likelihood serves as the metric in which "best-fitting" is defined in maximum likelihood estimation.

Logarithmic spacing. Spacing of numbers according to a geometric series, i.e., in which the ratios of adjacent pairs of numbers are the same. The ith value of a set of n logarithmically spaced values starting with a and ending with b is:

$$x(i) = 10^{[\log a + (i-1)\log(b/a)/(n-1)]}$$

Logarithmically spaced values can be computed in MATLAB® using:

```
>>X=logspace(log10(a),log10(b),n)
```

Log likelihood. Logarithmic transform (base e) of the likelihood (*see* Likelihood).

Logistic function.

$$F_L(x;\alpha,\beta) = \frac{1}{1 + \exp(-\beta(x - \alpha))}$$

where α determines the location (threshold) and β determines the slope of the function.

Luminance. Measure of light intensity. Common measures are candelas per square metre (cd/m^2) or foot-lamberts (fl or ft-L).

Magnitude estimation. Method for deriving a perceptual scale in which observers provide a numerical estimate of the perceived magnitudes of each stimulus magnitude.

Match-to-sample. Forced-choice procedure in which the observer views a "sample" stimulus and then selects the sample from a number of alternative "match" stimuli. The minimum number of stimuli is 3: one sample, two match.

Matrix. A two-dimensional array of numbers.

Maximum-likelihood estimation. Estimation procedure in which the best-fitting model is defined to be that which maximizes the likelihood function.

Maximum Likelihood Difference Scaling (MLDS). Method for deriving an interval perceptual scale from judgements about perceived stimulus differences, in which the perceptual values corresponding to each stimulus magnitude are estimated using a maximum likelihood criterion.

Metamers. Stimuli that are physically different yet perceptually indiscriminable.

Method of adjustment. Method in which observers freely adjust the magnitude of a stimulus in order to reach a criterion, for example a threshold or PSE.

Method of constants. Method in which the magnitude of the stimulus presented on each trial is selected from a predefined set.

Method of limits. Method in which observers are presented with a series of stimuli of either increasing (ascending method of limits) or decreasing (descending method of limits) magnitude, and report when the stimulus appears to change state, e.g., from visible to invisible or *vice versa*. A threshold is considered to be the stimulus magnitude at which the change of state occurs. Typically, the ascending and descending methods are used alternately and the thresholds from each are averaged, minimizing errors due to habituation and expectation.

Method of paired comparisons. Method for deriving a perceptual scale involving stimulus pairs. On each trial two stimuli are selected from a range of stimuli and the observer decides which has the higher perceived magnitude. The set of pair responses are used to derive estimates of the perceptual values corresponding to each stimulus magnitude.

Method of quadruples. Method for deriving an interval perceptual scale involving four stimuli per trial. The stimuli are presented in two pairs and the observer decides which pair is more perceptually similar (or more perceptually different). The set of quadruple responses are used to derive estimates of the perceptual values corresponding to each stimulus magnitude.

Method of triads. Method for deriving an interval perceptual scale involving three stimuli per trial. One of the stimuli is allocated as the target and the observer decides which of the two remaining stimuli is most perceptually similar (or different) to the target. The set of triad responses are used to derive estimates of the perceptual values corresponding to each stimulus magnitude.

Michelson contrast. Defined as $(L_{max} - L_{min})/(L_{max} + L_{min})$ where L_{max} and L_{min} are the maximum and minimum luminances. Michelson contrast is the favored metric of contrast for periodic stimuli such as sinusoidal gratings, but is also applicable to any stimulus defined by two luminance levels.

Monochromatic. Light composed of a single or very narrow band of wavelengths.

Muller–Lyer illusion. The illusory difference in length between a line with acute-angle fins at both ends and a line with obtuse-angle fins at both ends.

Multi-partition scaling. Also termed the "simultaneous solution," the partition scaling method in which the observer adjusts the magnitudes of a range of stimuli until they appear

at equal perceptual intervals. The first and last stimulus magnitudes in the range are usually non-adjustable anchor points.
Multiplicative noise. Internal noise that is proportional to stimulus magnitude.
Nanometer. Unit of light wavelength λ, usually abbreviated to nm (10^{-9} meter).
Noise distribution. Distribution of the relative probabilities of noise samples of different magnitude.
Normal distribution.

$$\phi(x) = \frac{1}{\sigma\sqrt{2\pi}} \exp\left[-\frac{(x-\mu)^2}{2\sigma^2}\right]$$

where μ is the mean of the distribution and σ the standard deviation.
Nulling procedure. The procedure whereby a stimulus whose percept has been altered along some stimulus dimension is returned to its original perceptual state via a change to some other stimulus dimension.
Odd-man-out task. *See* Oddity task.
Oddity task. Forced-choice task in which the observer is presented with a number of stimuli, all but one of which are the same, and chooses the stimulus that is different. The minimum number of stimuli is 3.
One up/two down. Adaptive (or staircase) method that targets 70.71% correct. Stimulus magnitude is increased after each incorrect response and decreased after two consecutive correct responses.
One up/three down. Adaptive (or staircase) method that targets 79.4% correct. Stimulus magnitude is increased after each incorrect response and decreased after three consecutive correct responses.
Ordinal scale. Perceptual scale in which stimuli are rank-ordered according to perceived magnitude.
Paired comparisons. *See* Method of paired comparisons.
Partition scaling. Method for deriving a perceptual scale that involves observers adjusting a stimulus to be perceptually midway between two fixed, or anchor, stimuli.
Pedestal. The baseline stimulus to which an increment or a decrement in stimulus magnitude is added.
Perceptual scale. The function describing the relationship between the perceived and physical magnitudes of a stimulus dimension. Examples are perceived contrast as a function of contrast, perceived velocity as a function of velocity, perceived depth as a function of retinal disparity.
Point of subjective equality (PSE). The physical magnitude of a stimulus at which it appears perceptually equal in magnitude to that of another stimulus. An example is a stimulus with, say, a contrast of 0.5 that appears to have the same contrast as a larger stimulus with, say, a contrast of 0.4.
Point of subjective alignment. The relative positions of two lines at which they appear aligned.
Posterior probability. Reflects a researcher's beliefs regarding the truth of a hypothesis taking into account prior beliefs as well as empirical data. *See also* Bayes' Theorem.

Posterior odds. Reflects a researcher's beliefs regarding the relative probabilities of two alternative models of some data taking into account prior beliefs as well as empirical data.

$$\textit{posterior odds} = \frac{p(M_1 \mid \mathbf{y})}{p(M_2 \mid \mathbf{y})} = BF\frac{p(M_1)}{p(M_2)},$$

where BF is the Bayes Factor and $(p(M_1)/p(M_2))$ is the prior odds.
Power function. $F(x;a,n) = ax^n$.
Precision. The inverse of the variability of a psychophysical measurement. The measure of variability may be the spread of the psychometric function or the standard deviation of a set of measurements.
Prior probability. Reflects a researcher's beliefs regarding the truth of a hypothesis prior to the collection of empirical data. *See also* Bayes' Theorem.
Prior odds. Reflects a researcher's beliefs regarding the relative probabilities of two alternative models of some data.

$$\textit{prior odds} = \frac{p(M_1)}{p(M_2)},$$

where $p(M_i)$ reflects the researcher's prior belief in model M_i in terms of a probability.
Probability density function. Function describing the relative probabilities of events. The function must be integrated to derive actual probabilities.
Progressive solution (in partition scaling). Partition scaling method in which the observer divides the perceptual distance between two anchor points into two subjectively equal parts by adjusting a third stimulus to be perceptually midway between the anchors, then divides the two subjectively equal parts into four using two new adjustable stimuli, then into eight, etc., until the required number of partitions has been reached.
Proportion correct. The proportion of trials in which the observer makes a correct response.
Proportion false alarms. The proportion of target-absent trials in which the observer responds that the target is present.
Proportion hits. The proportion of target-present trials in which the observer responds that the target is present.
Psi method. Adaptive method which optimizes the efficiency of estimation of the threshold, as well as the slope parameter of a PF. On each trial the stimulus magnitude is chosen that will lead to the lowest expected entropy across the posterior distribution defined across threshold and slope parameters.
Psychometric function. A function that describes the relationship between probabilities of observer responses and stimulus magnitude. The general form of the psychometric function is:

$$\psi(x;\alpha,\beta,\gamma,\lambda) = \gamma + (1 - \gamma - \lambda)F(x;\alpha, \beta)$$

where $F(x; \alpha, \beta)$ is the function with parameter α determining the x value at which the function reaches some criterion value (e.g., 0.5) and β determines the slope of the function. Parameters γ and λ are the guess and lapse rates, respectively. Commonly used functions are the Logistic, Cumulative Normal, Weibull, Gumbel, and Hyperbolic Secant.

Pulsed-pedestal. Procedure in which a pedestal stimulus and its increment (or decrement) are presented in synchrony.

Quadruples. *See* Method of quadruples.

QUEST. Adaptive method which can be considered a Bayesian version of the best PEST (*see* best PEST). After each response, the posterior distribution across possible threshold parameter values is determined from the prior distribution, which reflects the experimenter's assumptions about the threshold, and the likelihood function based on all preceding trials. The threshold estimate with the highest posterior probability serves as the stimulus magnitude for the subsequent trial.

Rayleigh match. A traditional tool for studying color vision and diagnosing color deficiency. Defined as the relative intensities of a mixture of red (say 679 nm) and green (say 545 nm) light required to match a monochromatic yellow (590 nm) light.

Ratio scale. A perceptual scale in which the ratio of scale values corresponds to the ratios of perceived magnitudes of the corresponding stimuli. A ratio scale can be rescaled by aX without loss of information where a is an arbitrary constant.

Recognition. Refers to experiments/tasks in which the observer names a stimulus from memory or selects a stimulus previously shown from a set of choices. Most often used to characterize experiments involving complex stimuli such as faces, animals, household objects, etc.

Reflectance. The proportion of incident light reflected by an object.

Reliability. The reproducibility of a psychophysical measurement.

Response bias. *See* Bias (of observer)

Retinal disparity. The horizontal or vertical difference between the angle subtended by an object to each eye with respect to fixation.

Receiver Operating Characteristic. The function that describes the change in performance with the observer's criterion, in terms of the relation between the proportion of hits and proportion of false alarms.

RMS (Root Mean Square) Contrast. Defined as $RMS = \sqrt{\frac{1}{n}\sum_{i=1}^{n}\left(\frac{x_i - \bar{x}}{\bar{x}}\right)^2} = \frac{SD_x}{\bar{x}}$, where n is the number of pixels in the image, x_i is the luminance value of pixel i, $\bar{x}$ is the mean luminance, and SD_x is the standard deviation of luminance values x. Contrast measure of choice for complex images.

Same-different task. Task in which the observer decides whether a pair of stimuli are the same or are different. In the 1AFC version, one of the pairs (Same or Different) is presented on a trial and the observer responds "same" or "different". In the 2AFC version, both pairs (Same and Different) are presented on a trial, and the observer responds "first" or "second" depending on the alternative/interval perceived to contain the Same (or the Different) pair.

Sampling distribution. Probability density function of a statistic (e.g., parameter estimate). May be approximated by repeated estimation based on samples from an assumed population.

Sampling error. The difference between a parameter estimate and the parameter's true value.

Saturated model. A model that makes no assumptions other than the assumptions of stability and independence. As such, a saturated model contains a parameter for each unique stimulus condition. A model comparison that compares a more restricted model to the saturated model is known as a goodness-of-fit test.

Scalar. A single number, e.g., 8, 1.5, 1.4e-5.

Sensory scale. *See* Perceptual scale.

Signal distribution. Distribution of the relative probabilities of signal samples of various magnitude.

Signal Detection Theory. A theory of how observers make perceptual decisions based on the premise that the internal representation of a stimulus magnitude is a sampling distribution with a mean and a variance.

Sine-wave pattern. A pattern in which the stimulus dimension is modulated in space or time according to a sinusoidal function:

$$F(x;m,a,f,\rho) = m + a\ \sin(2\pi fx + \rho)$$

where m is the mean stimulus magnitude, a the amplitude of modulation, f the frequency of modulation (in cycles per unit space or time), and ρ the phase of modulation (in radians; one full cycle equals 2π radians). The inclusion of 2π in the equation means that a full cycle of modulation will be completed in the interval $0 < x < 1$.

Simultaneous brightness contrast. The phenomenon in which the brightness of a stimulus depends reciprocally on the luminance of its surround.

Simultaneous solution (in partition scaling). *See* Multi-partition scaling.

Slope (of psychometric function). Rate of change of response as a function of stimulus magnitude. One of the four parameters that characterize a PF (β). Note, however, that whereas β is often referred to as the slope of the PF, it generally will not correspond in value to the slope of the function as defined in calculus (i.e., the first derivative of the function).

Spatial frequency (SF). The number of cycles of modulation of a stimulus dimension per unit visual angle. Typically measured as cycles per degree (cpd).

Spread (of psychometric function). Also known as support of psychometric function. Stimulus range within which a PF goes from $\gamma + \delta$ to $1 - \lambda - \delta$, where γ is the lower and $1 - \lambda$ the upper asymptote of the PF. δ is an arbitrary constant ($0 < \delta < [1 - \lambda - \gamma]/2$). Thus, if we let σ symbolize spread:

$$\sigma = \psi^{-1}(1 - \lambda - \gamma;\alpha,\beta,\gamma,\lambda) - \psi^{-1}(\gamma + \delta;\alpha,\beta,\gamma,\lambda)$$

where $\psi^{-1}(y;\alpha,\beta,\gamma,\lambda)$ is the inverse of the psychometric function $\psi(x;\alpha,\beta,\gamma,\lambda)$

Stability, assumption of. The assumption that the performance of an observer (for example, the probability of a correct response as a function of stimulus intensity x) does not change during the course of the experiment.

Staircase methods. *See* Adaptive methods.

Standard deviation. A measure of the variability among scores. For any set of numbers x_i, the standard deviation is given as:

$$\sigma = \sqrt{\frac{\sum_{i=i}^{n}(x_i - \overline{x})^2}{n}},$$

where $\overline{x}$ is the mean of x, and n is the number of scores. If the numbers x_i are a random sample drawn from a population, the following expression is that of an unbiased estimate of the standard deviation of the population from which the xs were drawn.

$$\hat{\sigma} = \sqrt{\frac{\sum_{i=i}^{n}(x_i - \overline{x})^2}{n - 1}}$$

Standard error. The standard deviation of a parameter's sampling distribution. Used to quantify the reliability of a parameter estimate.

Standardized normal distribution. The normal distribution with mean equal to 0 and standard deviation equal to 1.

$$\phi(z) = \frac{1}{\sqrt{2\pi}}\exp\left(\frac{-z^2}{2}\right)$$

Steady-pedestal. Procedure in which the pedestal stimulus is presented alone before the addition of the increment or decrement.

Stimulus exposure duration. The length of time a stimulus is exposed during a trial.

Stimulus-onset-asynchrony (SOA). The temporal interval between the onset of two stimuli.

Stereopsis. The means by which the relative depth of an object is determined by virtue of the fact that the two eyes view the object from a slightly different angle.

Support (of psychometric function). *See* Spread of psychometric function.

Symmetric (form of 1AFC). Type of single-alternative forced-choice task/procedure in which the two discriminands can be considered to be mirror opposites, for example grating patches that are left- and right-oblique.

Temporal frequency (TF). The number of cycles of modulation of a stimulus dimension per unit time. Typically measured as cycles per second (cps).

Termination criterion. In adaptive methods, the rule that is used to terminate a staircase. For example, a staircase may be terminated after a set number of trials or a set number of reversals.

Thurstonian scaling. Method for deriving an interval perceptual scale using the method of paired comparisons, in which the scale is derived from the proportions of times that each stimulus magnitude is perceived to be greater than each of the other stimulus magnitudes.

Threshold. In general refers to the difference in magnitude between two stimuli or stimulus states that enables them to be just discriminable. Examples are a contrast detection threshold, a contrast discrimination threshold, or the threshold for binocular rivalry.

Threshold-versus-contrast (TvC). The function relating the threshold for detecting an increment (or decrement) in contrast as a function of the pedestal (or baseline) contrast.

Threshold-versus-intensity (TvI). The function relating the threshold for detecting an increment (or decrement) in intensity (or luminance) as a function of the pedestal (or baseline) intensity.

Two-alternative forced-choice (2AFC). Here defined as any procedure in which the observer selects a stimulus from two alternatives. Examples are selecting the left oblique grating from a left- and a right- oblique grating pair, or choosing from two alternatives a stimulus previously shown.

Transducer function. *See* Perceptual scale.

Transformed Likelihood Ratio (TLR). Statistic used to determine whether two models differ significantly. When one of the two models is nested under the other, *TLR* is asymptotically distributed as χ^2 with degrees of freedom equal to the difference in the number of free parameters between the two models. When the fuller model is the saturated model, the transformed likelihood ratio is known as deviance.

Triads. *See* Method of triads.

Triangular method. Alternative name for a 3AFC oddity task.

Two-interval forced-choice (2IFC). Procedure in which the observer selects a stimulus from two stimuli presented in a temporal sequence.

Type 1 experiment. A psychophysical experiment/procedure/task in which there is a correct and an incorrect response on each trial.

Type 2 experiment. A psychophysical experiment/procedure/task in which there is no correct and incorrect response on each trial.

Vector. An $m \times 1$ or $1 \times n$ array of numbers.

Vernier acuity. The smallest misalignment of two stimuli that can be reliably detected.

Vernier alignment. Experiment/task aimed at measuring the threshold (or precision) for detecting that two stimuli are misaligned, and/or measuring the physical separation at which the two stimuli are perceived to be aligned, i.e. the bias.

Visual acuity. Measure of the acuteness or clearness of vision. Traditionally measured using an eye chart.

Visual angle. The angle subtended by a stimulus to the eye. Usually measured in arc degrees, arc minutes or arc seconds.

Weber contrast. Defined as $(\Delta L / L_b)$ where ΔL is the difference between the luminance of the stimulus and its background, and L_b is the luminance of the background. Weber contrast is normally employed to measure the contrast of a uniform patch on a background, and is not normally used for periodic stimuli or noise patterns.

Weber's Law. Law that states that the just discriminable difference in stimulus magnitude is proportional to stimulus magnitude.
Weibull function.

$$F_W(x;\alpha,\beta) = 1 - \exp\left(-\left(\frac{x}{\alpha}\right)^{\beta}\right)$$

where α determines the location (threshold) and β determines the slope of the function.
Yes/No. Experiment/task in which a single stimulus is presented on each trial and the observer is required to indicate whether or not it contains the target.
z-score. A score that corresponds to the number of standard deviations a score is above (if positive) or below (if negative) the mean. The *z*-scores corresponding to any distribution of scores will have a mean equal to 0 and a standard deviation equal to 1. The *z*-scores corresponding to any normally distributed variable will be distributed as the standard normal distribution.

List of Acronyms

AFC.	Alternative-forced-choice
AIC.	Akaike's information criterion
APE.	Adaptive probit estimation
BIC.	Bayesian information criterion
CPD.	Cycles per degree
CPS.	Cycles per second
CRT.	Cathode ray tube
IFC.	Interval-forced-choice
ISI.	Inter-stimulus-interval
ITI.	Inter-trial-interval
JND.	Just-noticeable-difference
LL.	Log likelihood
LR.	Likelihood ratio
M-AFC.	M-alternative-forced-choice
MDS.	Multi-dimensional scaling
ML.	Maximum likelihood
MLDS.	Maximum likelihood difference scaling
1AFC.	One-alternative-forced-choice
1IFC.	One-interval-forced-choice
PEST.	Parameter Estimation by Sequential Testing.
PF.	Psychometric function
PSA.	Point-of-subjective-alignment
PSE.	Point-of-subjective-equality
ROC.	Receiver operating characteristic
RT.	Reaction time
SD.	Standard deviation
SDT.	Signal detection theory

SE.	Standard error
SOA.	Stimulus-onset-asynchrony
SF.	Spatial frequency
TLR.	Transformed likelihood ratio
TF.	Temporal frequency
TvC.	Threshold versus contrast
TvI.	Threshold versus intensity
2AFC.	Two-alternative-forced-choice
2IFC.	Two-interval-forced-choice
3AFC.	Three-alternative-forced-choice

Index

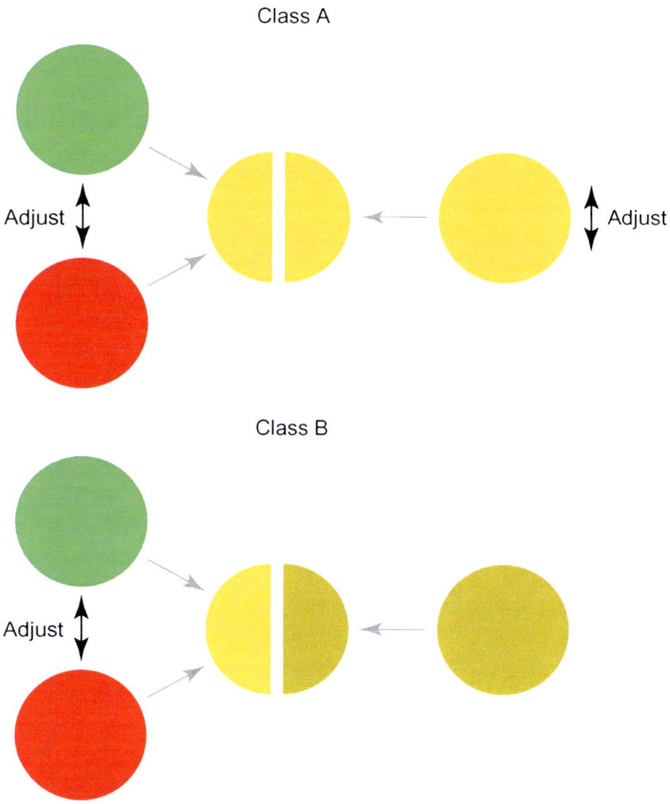

FIGURE 2.3 The Rayleigh match illustrates the difference between a Class A and Class B psychophysical observation. For Class A, the observer adjusts both the intensity of the yellow light in the right half of the bipartite field, as well as the relative intensities of the red and green mixture in the left half of the bipartite field, until the two halves appear identical. For Class B, the observer adjusts only the relative intensities of the red and green lights to match the hue of a yellow light that is different in brightness.

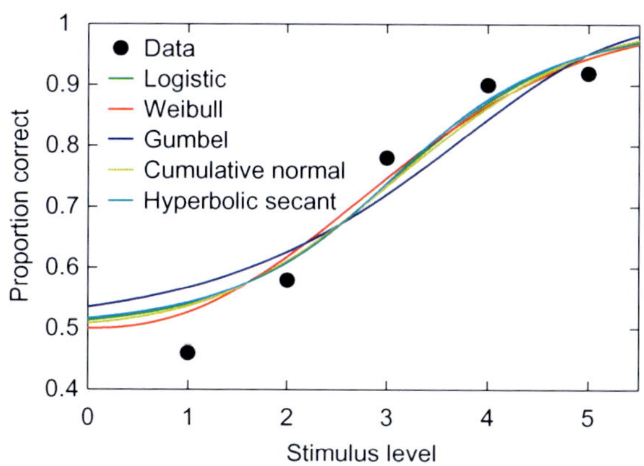

FIGURE 4.3 Example fits of five different PF functions.